The Evolution of Hazardous Waste Programs: Lessons from Eight Countries

by Katherine N. Probst
and Thomas C. Beierle

Center for Risk Management • Resources for the Future • June 1999

Published 1999 by Resources for the Future

Published 2017 by Routledge
2 Park Square, Milton Park, Abingdon, Oxon, OX14 4RN
711 Third Avenue, New York, NY 10017

First issued in hardback 2017

Routledge is an imprint of the Taylor & Francis Group, an informa business

Library of Congress Cataloging-in-Publication Data

Probst, Katherine N.
    The evolution of hazardous waste programs : lessons from eight countries / by Katherine N. Probst and Thomas C. Beierle.
        p.   cm.
    Includes bibliographical references (p.).
    ISBN 0–891853–01–5 (pbk.)
    1. Hazardous wastes—Management. 2. Hazardous wastes—Law and legislation. 3. Hazardous waste treatment facilities—Finance. I. Beierle, Thomas C. II. Title.
TD1030.P75   1999
363.72'8756—dc21   99-21298
                                                                                       CIP

This book is a product of the Center for Risk Management at Resources for the Future, J. Clarence Davies, director. Diane Kelly, Kelly Design, designed the book and its cover, and typeset it in ITC Berkeley.

ISBN 13: 978-1-138-40710-7 (hbk)
ISBN 13: 978-1-891853-01-2 (pbk)

Contents

Acknowledgments

The research in this report was funded jointly by the World Bank's South Asia Environment Group and by general support from Resources for the Future (RFF).

The impetus for this research came from many conversations the authors had with managers and staff members at the World Bank, especially with Richard Ackermann, Bekir Onursal, and Michele Keene. Both Mr. Ackermann and Mr. Onursal provided useful ideas throughout our work, as well as financial support. We appreciate their insights and their willingness to help fund this research.

We wish to thank the many people—too numerous to list here—who provided information about hazardous waste programs and facility financing in the countries studied and who provided constructive criticisms of draft versions of this report and the country profiles. A few of these people deserve special mention, as they provided help above and beyond the call of duty. They include Chris Clarke, former editor of the *Financial Times' Environmental Liability Report* in England; Matt Hale of the U.S. Environmental Protection Agency; Dan Millison of Ecology and Environment in the United States; Sombat Sae-Hae of the Thailand Development Research Institute; and Tom Walton of the World Bank.

We could not have issued this report without the help of a number of people at RFF. Alan Krupnick and Ruth Bell provided wise counsel and good ideas throughout the project. Terry Davies provided useful advice, as always. Chris Kelaher and Betsy Kulamer provided their usual high-caliber publications assistance.

The views expressed in this report are those of the authors and should not be ascribed to the persons or organizations whose assistance is acknowledged above or to the trustees, officers, or other staff members of Resources for the Future.

Executive Summary

In most countries, the development of environmental programs follows a similar pattern. Early efforts concentrate on direct threats to public health, such as contaminated drinking water and air pollution. Only after these problems are addressed does the issue of improving the day-to-day management of wastes deemed "hazardous" rise to the top of the nation's environmental agenda.

In the past thirty years, many developed countries have established effective hazardous waste management programs. During the past decade, some developing countries—particularly those that have experienced rapid economic growth and industrialization—have begun to consider ways of developing and implementing programs to assure the proper disposal of hazardous waste. No two countries share identical circumstances in terms of political regime, industrial policy, major industries, geography, and the nature of their hazardous waste problem; an examination of the experiences of both developed and developing countries offers useful insights about the evolution of hazardous waste management programs to other countries embarking on a similar path.

Managing Hazardous Waste: A Difficult Challenge

Fundamentally, the goal of an effective hazardous waste management program is changing the behavior of those organizations (both public and private) that generate and manage hazardous wastes. The key components of meeting this goal are, first, building an effective regulatory program and, second, developing adequate treatment, storage, and disposal facilities. Both components present challenges to a country seeking to move from a situation in which there is little or no regulation of hazardous waste to one in which the majority of generators treat, store, and dispose of hazardous waste in an environmentally safe way.

In this report, we examine the lessons that can be learned from eight countries that have instituted (or have begun to institute) hazardous waste management programs during the last thirty years. We look at the experiences of four developed countries—Germany, Denmark, the United States, and Canada—and four developing countries—Malaysia, Hong Kong, Thailand, and Indonesia. Each of these (and in that order) have tackled the hazardous waste management challenge. This report represents a first effort at a cross-country comparison of the evolution of hazardous waste management programs around the world. We look at a small sample of countries, from three regions of the globe. However, the countries studied employ a broad range of approaches, and their experiences highlight a number of important issues.

In many ways, the central challenge of developing a successful hazardous waste management program is creating incentives for proper hazardous waste treatment and disposal. To compel compliance, hazardous waste management programs typically rely on what can be called "negative incentives"—requirements that impose costs on the waste generator and carry with them the "stick" of permits, licenses, inspections, and enforcement. Many countries have chosen to make it easier to comply in the early years of a new regulatory regime by "ramping up" the stringency—and often the costs—of requirements over time. The first issue we address in this report is: *What are the major steps in the evolution of a successful hazardous waste management program, and how long does it take before new requirements result in improvements in the way that hazardous wastes are actually managed?*

Assuring adequate facilities for treating, storing, and disposing of hazardous wastes presents its own challenges. There are usually few high-quality waste management facilities in countries that are just beginning to institute hazardous waste management requirements. And, in the early years of a program, it is difficult to attract private-sector money to build needed facilities. Often, private investors cannot be assured that hazardous waste generators will be willing to pay the higher costs of proper waste treatment and disposal, rather than dumping hazardous wastes in back lots and rivers. When this is the case, some form of public-sector financing often is used to get needed treatment and disposal infrastructure on-line. Public-sector financing also allows governments to introduce the positive incentive of subsidized disposal fees that encourage waste generators to use high-quality waste management facilities at low (or no) cost. The second key question we address is: *What role, if any, does the public sector play in financing hazardous waste treatment and disposal facilities?*

Ultimately, the issues of regulatory program development and facility financing are inextricably linked. Absent a set of clear rules and adequate enforcement, industries have little incentive to pay for proper waste disposal;

this leaves government and private firms with little financial incentive to invest in expensive waste treatment and disposal. Without adequate facilities, however, it is very difficult (if not impossible) to hold the regulated community accountable for proper waste management.

Creating an Effective Hazardous Waste Management Program

The experiences of the eight countries in our study make one lesson clear: It takes a long time to develop an effective hazardous waste management program. Programs evolve through a complex process subject to the particular economic, political, legal, and cultural context of individual countries. As programs evolve, however, they typically pass through five major stages.

1. *Identifying the problem and enacting legislation*: recognizing that an environmental problem exists and enacting legislation to address it.
2. *Designating a lead agency*: giving authority to a specific agency, or agencies, to draft, implement, and enforce regulations.
3. *Promulgating rules and regulations*: establishing the legal basis for a regulatory program, including (a) identifying which wastes will be subject to regulation and (b) identifying specific technical, procedural, and information requirements for waste treatment, storage, and disposal facilities and for generators of hazardous wastes.
4. *Developing treatment and disposal capacity*: ensuring the construction and operation of hazardous waste management facilities using public funds, private investment, or a combination of the two.
5. *Creating a mature compliance and enforcement program*: influencing the behavior of generators and operators of hazardous waste management facilities to ensure that waste is properly managed. At this final stage, one can say that a "culture of compliance" exists. Once a program has matured to this point, renewed effort is usually focused on ways to reduce or recycle hazardous substances to decrease the need for expensive waste treatment and disposal.

Each of these stages takes a number of years, and, at each stage, there are many difficult issues to be resolved. Germany, Denmark, the United States, and Canada began the process of program development during the 1970s. For the most part, their regulatory programs were fully operational by the end of the 1980s, and subsequent laws and policies have focused mainly on encouraging waste minimization and recycling. Malaysia, Thailand, Hong Kong, and Indonesia began focusing serious attention on hazardous waste management in the late 1980s and early 1990s (although some began initial efforts in the early 1980s). With the possible exception of Hong Kong, these developing countries are still some way from having what most would describe as fully operational programs. Table ES-1 presents a summary of the dates of each country's first major hazardous waste management laws or regulations.

A review of the evolution of hazardous waste management programs in the eight countries we examined suggests some general lessons that should prove useful to those countries contemplating the creation of their own hazardous waste management programs. Perhaps the most significant lesson, however, is one of humility about the enormity of the task.

1. It takes a long time—at least ten to fifteen years—to develop a fully operational hazardous waste regulatory system.

It took the United States, Germany, Canada, and Denmark ten to fifteen years to develop the laws, institutions, and procedures that resulted in widespread changes in the way firms handle, treat, and dispose of hazardous waste. These are all "rule-of-law" countries with strong legal and bureaucratic institutions. Hong Kong, Indonesia, Malaysia, and Thailand have been actively regulating hazardous waste for five to ten years and still have a long way to go before hazardous waste is properly managed in a comprehensive way.

Table ES-1. Dates of First Major Laws or Regulations Dealing with Hazardous Waste Management

Developed Countries	
Germany	1972 (law)
Denmark	1973 (laws)
United States	1976 (law)
Canada	1980 (law)
Developing Countries	
Malaysia	1989 (regulations)
Hong Kong	1991 (laws)
Thailand	1992 (laws)
Indonesia	1994 and 1995 (regulations)

2. Developing a culture of compliance is the crucial element of an effective hazardous waste management system.
No government has enough resources to inspect every hazardous waste management facility frequently enough to detect all violations of rules and regulations. Instead, the regulatory system must have enough public credibility—often established by making the threat of enforcement real—that most regulated entities comply with environmental requirements as a matter of course. Absent such a culture, it is extremely difficult to have an effective hazardous waste management program. The question of what, if anything, can be done to *create* a culture of compliance is far larger than simply building effective environmental programs. It raises fundamental issues of governance and the core legal and political culture of a country.

3. Clear lines of regulatory authority increase the chances of successful implementation of new regulatory programs.
A single *environmental* agency may not always be the best way to proceed, but having clear lines of regulatory authority vested in a single agency is important to program success. Simply having good laws and regulations on the books does not guarantee that a hazardous waste management program will be successful. What is needed are government institutions that are credible and that have the resources and expertise to implement regulatory requirements.

4. There are important consequences to the decision *not* to harmonize policy at the national level.
While most of the countries we studied had some form of shared responsibility between the national government and subnational jurisdictions, some delegated most of the responsibility for the design and implementation of programs to provinces and states. In these cases, decentralization resulted in differences in the quality and effectiveness of different state- and province-level programs, as well as differences in waste definitions, permitting requirements, and treatment and disposal standards. These countries have subsequently sought greater harmonization of their hazardous waste policies at the national level.

Bringing Needed Hazardous Waste Management Facilities On-Line

Ideally, once a regulatory system has been implemented, there would also be adequate capacity for the proper treatment and disposal of hazardous waste.

But hazardous waste management facilities do not come on-line overnight. They are expensive, and the market for hazardous waste treatment and disposal is highly uncertain, especially in the early years of a regulatory program. Often, private investors are unsure whether hazardous waste generators will be willing to pay disposal fees that are high enough to provide an adequate return on investment. This is especially true in countries with weak legal and enforcement regimes.

In many countries, in order to assure that needed hazardous waste management facilities *do* come on-line, the public sector has provided direct financial support to build and operate facilities or has offered incentives that reduce private investors' financial risk. In some cases, government agencies build and operate hazardous waste management facilities themselves.

Partial or full public financing of hazardous waste facilities creates the option of subsidizing disposal fees. Subsidies are an important policy tool, as they help to discourage illegal disposal by making proper hazardous waste management cheaper. Subsidies can be available to all, or they can be targeted to some generators, such as small businesses, with the least ability to pay for disposal. While subsidies can be effective in helping to control illegal disposal, they may also work against another important, long-term, hazardous waste management goal—getting generators to minimize their production of hazardous waste in the first place. However, preventing illegal disposal is typically a more urgent concern in the early years of regulatory programs. Attention to minimizing hazardous waste generation often comes later, as regulatory programs mature.

The countries we studied ran the gamut from a private-sector approach in the United States, Malaysia, and Indonesia to a public-sector approach in Denmark and parts of Germany. The remaining countries fell somewhere between, with shared public and private sector responsibilities. Table ES-2 presents a summary of the public- and private-sector financing approaches employed by the countries examined in this study.

Based on our examination of eight countries, we reached three important conclusions about facility financing.

1. There is no single "proper" approach to hazardous waste management facility financing that will work in every country.
No one financing approach—private, public, or a mix—was clearly superior to the others in all cases. Rather than adopting a standard approach, each country

Table ES-2. Summary of Public- and Private-Sector Financing of Hazardous Waste Management Facilities

Developed Countries	
Germany	public/private
Denmark	public
United States	private
Canada	public/private
Developing Countries	
Malaysia	private
Hong Kong	public/private
Thailand	public/private
Indonesia	private

should select a financing approach that is tailored to its circumstances (industrial profile, geography, government resources, and capacity), the effectiveness of its regulatory system, and its general policy objectives.

2. In countries where there is not yet a culture of compliance, the financing approach matters.

Although there is no standard approach to financing hazardous waste facilities, it is clear that where enforcement and compliance are weak—a situation typical of the early years of regulatory programs—subsidies are an important policy tool for encouraging generators to properly manage hazardous wastes. In this environment, treatment and disposal facilities usually "compete with the river," where the cost of disposal is zero. When the regulatory stick is weak—as in all of the developing countries we examined—financing models that allow countries to offer the carrot of subsidized disposal fees are likely to be more effective than other approaches in encouraging a change in behavior.

3. Disposal fee subsidies are a viable transitional strategy for encouraging proper waste disposal.

Several countries we studied showed that transitions from a subsidized approach toward a more market-driven approach are possible. Subsidies can encourage early compliance, and getting generators in the habit of using hazardous waste management facilities is an important first step toward the long-term goal of building a culture of compliance. More stringent and costly hazardous waste management requirements can be phased in over time.

Other Important Issues and Areas for Further Research

In the course of our research, several issues arose that are not directly related to program evolution or facility financing but are, nonetheless, important elements in developing a hazardous waste management program. We briefly describe them here and then recommend issues warranting additional research.

Other Important Issues

1. One very important resource in a hazardous waste management program is *information*—about who is generating waste, what quantities and types are being generated, and where it is going.

2. Regardless of whether facilities are financed publicly or privately, accurately estimating needed future capacity is difficult.

3. Planning for hazardous waste infrastructure must account for the geography of hazardous waste generation and the cost of transportation from generators to treatment and disposal facilities.

4. There are a number of nonmarket approaches that can be (and have been) used to encourage waste reduction and recycling, including laws and policies to encourage waste minimization, public information, and subsidies or requirements for waste reduction equipment.

5. Siting hazardous waste facilities is always controversial.

Areas for Further Research

Several important issues warrant further study. Most important is the question of what can be done to create a culture of compliance. It is clear that such a culture plays an important role in hazardous waste management and other areas of environmental policy, yet the steps to building such a culture are poorly understood.

Also poorly understood are the specific elements of two financing issues. First, how should subsidies be structured to "get the incentives right" for hazardous waste generators, investors, and the government? And, second, given that public–private ventures are so prevalent in hazardous waste management facilities, how should such arrangements be structured in different situations?

Finally, our research focused on eight countries in three regions of the globe. What have been the experiences in other parts of the world, such as Eastern Europe, Latin America, and Africa? It would be useful to learn whether the findings for countries in these regions are consistent with those reached in this report.

1. Introduction

In most countries, the development of environmental programs follows a similar pattern. Early efforts to clean up the environment focus on immediate and direct threats to public health, protecting drinking water and reducing air pollution. Typically, the focus on hazardous waste management comes as part of the second phase of a country's environmental programs, after more urgent problems have been addressed.

For understandable reasons, the development of environmental programs in developing countries lags behind that of developed countries by a decade or two. While developed countries, such as the United States, Denmark, Canada, and Germany, initiated comprehensive hazardous waste management programs in the late 1970s and early 1980s, developing countries, for the most part, did not begin to turn their attention to this set of environmental problems until the 1990s.

The road to establishing an effective hazardous waste management (HWM) program is a bumpy one, even in developed countries. It involves changing the behavior of those who generate hazardous wastes so that they routinely store, transport, treat, and dispose of hazardous waste in an environmentally safe manner. An examination of the experiences of both developed and developing countries offers some important lessons in terms of two critical elements: first, designing an effective regulatory program aimed at changing the behavior of

1

waste generators and, second, creating a market for facilities that properly treat and dispose of hazardous waste.

In many ways, the central challenge to developing a successful HWM program is creating incentives for proper hazardous waste treatment and disposal. The demand for HWM services is driven largely by regulation, whether it is a command-and-control type or a more market-based approach. Generators of hazardous waste must have incentives (of either the "carrot" or "stick" variety) to treat and dispose of waste properly and to pay the associated costs. One of the concerns about tightening up the regulatory system is that it makes proper waste treatment and disposal more expensive, creating a powerful incentive for *improper* disposal. In the early years of many countries' HWM programs, proper hazardous waste treatment and disposal "competed with the river," with the cost of disposal being effectively zero. It takes time to build a regulatory system that can discourage such low-cost and environmentally harmful waste management practices. A key question, which forms a major aspect of our research, is: How long does it take to progress through the various stages of developing an HWM program until one can say that a "culture of compliance" exists among generators and that proper waste management is the norm, rather than the exception?

Ideally, once a regulatory system has been implemented, there would be adequate capacity for the proper treatment and disposal of hazardous waste. However, waste management facilities do not come on-line overnight. Typically, some companies build hazardous waste treatment, storage, and disposal (TSD) facilities on site to manage their own wastes. Other TSD facilities are built off site and are available on a fee-for-service basis. We refer to these off-site HWM facilities as "commercial" facilities, and they (not the on-site facilities) are the focus of this report.

Commercial TSD facilities are costly to build, and siting is usually quite controversial. Thus, ensuring adequate capacity is not an easy task. Often, there is little incentive for private investment in waste treatment infrastructure until there is a mature regulatory system. Without such a system, it is difficult to accurately predict demand in what is essentially a new market. To what extent will generators comply with new rules? Will they meet regulatory requirements by reducing waste generation rather than paying for new and expensive treatment and disposal? How much illegal disposal will remain? All of these factors, and the fact that waste treatment facilities can be very capital intensive, conspire to discourage investment in HWM infrastructure in the early years of a regulatory program. Governments can reduce some of this uncertainty by making proper treatment and disposal less expensive through subsidies. Yet, there is a delicate balance between lowering the price of waste disposal and still providing incentives for waste reduction. Finally, it is often hard to get the "right" facilities online. Predictions of demand must address such issues as the type of waste being generated and its geographical location. Not all treatment facilities are appropriate for all types of waste, and high transportation costs can greatly discourage the use of distant HWM facilities.

Research Approach

The purpose of this report is to examine the experiences of a small sample of developed and developing countries that have implemented HWM programs and to identify which approaches have worked and why. In examining the effectiveness of various approaches to hazardous waste management, we focus on basic programmatic definitions of "success": first, whether hazardous waste treatment and disposal capacity is available and, second, whether generators actually use these facilities. In short, our definition of an effective HWM program is a situation in which environmentally sound HWM practices are the norm. In this report, we present an overview of the ways in which the eight countries we examined developed an HWM program, and then we identify "lessons learned" from their experiences. Taken together, we hope these lessons can help developing countries design and implement successful HWM programs in the future.

We surveyed the development and implementation of hazardous waste management programs in eight countries: four developed countries—the United States, Canada, Germany, and Denmark; and four developing countries in Southeast Asia—Indonesia, Thailand, Malaysia, and Hong Kong* For simplicity, Hong Kong, a special administrative region of the People's Republic of China, is referred to as a "country" throughout this report, since most of our discussion reflects the period prior to Hong Kong's reversion to Chinese sover-

---

*The World Bank South Asia staff identified these countries as including a range of approaches and experiences. Because significant aspects of hazardous waste management programs in Canada and Germany are under provincial and state jurisdictions, we chose to focus on the provinces of Quebec and Alberta in Canada and on the states of Hesse and Bavaria in Germany.

eignty in 1997. While many aspects of Hong Kong's economy would suggest it could be categorized as a developed country, we have grouped it with the developing countries, based on the status of its hazardous waste program.

For each of the eight countries, we examined the evolution of its HWM program, identified the different approaches taken to developing commercial HWM facilities, and sought to formulate a picture of the steps taken to "create a market" for proper hazardous waste management. We collected information from secondary sources and interviewed in-country experts for each country. This report presents what we learned from the experiences of these countries—about the practices that have led to effective government programs, the mix of incentives and subsidies that helped create needed facilities, and the consequences of different approaches. We describe a range of approaches and identify those that have been successful and those that have not—and if not, why not.

To document the evolution of the countries' regulatory programs, we charted the achievement of certain key milestones, such as passing a hazardous waste law, implementing regulations, and developing treatment and disposal capacity. In addition, we collected information on the makeup of the hazardous waste market; the environmental regulatory institutional structure; the allocation of responsibilities among federal, state (or provincial), and local governments; and the overall effectiveness of the regulatory system.

To address questions about infrastructure financing, we collected information on public and private financing, ownership, and operation of commercial HWM facilities. We paid special attention to the degree to which disposal fees were subsidized (and whether subsidies changed over time) and to what, if any, incentives were provided by governments to encourage private investment. We also collected available information on the total amount of hazardous waste generated, the type and capacity of major commercial HWM facilities, and their financial performance and viability.

Following our review of secondary sources and expert interviews, we compiled the information into eight country profiles, which were then sent to reviewers in each country for comment and correction. These country profiles are included in Appendix A. Appendix B lists the names of those who provided information on the programs in each country, those who reviewed the individual country profiles, and those who reviewed a preliminary draft of the complete report.

Caveats

Surveying the experience of eight different countries reflects a choice of breadth over depth. Developing an HWM system is a complicated undertaking, and the path taken often depends on the particular geography, demographics, industrial profile, politics, and culture of a country. When possible, we discuss these subtleties. Although there are clearly major differences among countries, each provides a real-life experiment from which general lessons on the success and failure of different approaches can be derived. An in-depth examination of the experience of one or a few countries would, however, allow a more contextual and data-rich evaluation. The "lessons learned" described in this report provide a starting point for such an examination.

Our report focuses on the experiences of countries in Southeast Asia, Europe, and North America. Although it covers a wide range of approaches to hazardous waste management, it is wanting in terms of geographic diversity. Research on the experiences of countries in Eastern Europe, Latin America, and Africa would provide additional insights into the issues discussed here.

A number of issues that influence the environment in which regulatory systems and hazardous waste markets function are not covered in this report. Most obvious is recycling and reuse of hazardous waste—an issue of increasing importance. Other issues not addressed include the transboundary movement of hazardous waste (and legal restrictions on such movement) and the siting of hazardous waste facilities. Although these issues are important to a full understanding of hazardous waste management, each deserves in-depth treatment in its own right.

Outline of Report

In Chapter 2, we begin with a description of the evolution of HWM programs generally and then describe the experiences of the countries in our sample. Chapter 2 concludes with lessons learned regarding the successes and difficulties of trying to create an effective regulatory regime. In Chapter 3, we focus on public- and private-sector roles in financing and operating HWM facilities and again conclude with lessons learned based on the experiences of the eight countries in our study. Finally, in Chapter 4, we briefly describe a number of important issues for developing an effective hazardous waste management system that are not directly related to program development and facility financing. We conclude with suggestions of important areas that warrant additional research. Appendix A includes a brief description of the issues covered in this report on a country-by-country basis.

2. Hazardous Waste Program Development: A Multistage Process

The experiences of the eight countries in our study make one lesson clear: It takes a long time to develop an effective hazardous waste regulatory program. In the years preceding the development of such a program, uncontrolled disposal of hazardous waste is the norm; few, if any, state-of-the-art treatment and disposal facilities exist; and companies have few incentives to minimize waste. Information on who is generating waste, what types are being generated, and where wastes are being disposed of is meager or nonexistent.

The goal of a hazardous waste management (HWM) program is to make the transition to a situation in which environmentally responsible waste management is the norm and uncontrolled disposal is the exception. In an ideal environment, hazardous waste would be managed in licensed facilities, and generators would pay the full cost of disposal and have incentives to minimize the use and disposal of hazardous substances. Regulators would have accurate information on waste generation and disposal, and the public would have some degree of confidence in the effectiveness of the regulatory system.

The move from an unregulated environment to this vision of environmentally responsible waste management is a complex one. HWM programs typically evolve through a number of major stages, including enacting legislation, promulgating rules and regulations, and permitting (or licensing) treatment and disposal facilities.

In this chapter, we first describe the major stages in the evolution of environmental programs generally, with a focus on hazardous waste. The timing of these stages may differ by country, and stages may overlap, but each country we examined went (or is going) through a similar evolutionary process. We then summarize the overall evolution of HWM programs in the eight countries in our study. We conclude this chapter with lessons learned regarding program development.

Major Stages of Program Development

Typically, there are five major stages in the development of an HWM program:

1. *Identifying the problem and enacting legislation*: recognizing that an environmental problem exists and enacting legislation to address it.
2. *Designating a lead agency*: giving authority to a specific agency, or agencies, to draft, implement, and enforce regulations.
3. *Promulgating rules and regulations*: establishing the legal basis for a regulatory program, including (a) identifying which wastes will be subject to regulation and (b) identifying specific technical, procedural, and information requirements for waste treatment, storage, and disposal facilities and for generators of hazardous wastes.
4. *Developing treatment and disposal capacity*: ensuring the construction and operation of HWM facilities using public funds, private investment, or a combination of the two.
5. *Creating a mature compliance and enforcement program*: influencing the behavior of generators and operators of HWM facilities to ensure that waste is properly managed. At this final stage, one can say that a culture of compliance exists. Once a program has matured to this point, renewed effort is usually focused on how to reduce or recycle hazardous substances to decrease the need for expensive waste treatment and disposal.

Although the development of HWM programs has followed a similar progression in each country, there are a number of important differences among them. Different countries face different local circumstances, and the choices they make are often unique to those individual countries. The following discussion describes each of the five development stages and the various approaches taken in the different countries. Time lines charting the evolution of the hazardous waste program in each of the eight countries studied are included in the country profiles in Appendix A.

Identifying the Problem and Enacting Legislation

The first stage in developing an HWM program is the recognition that current HWM practices are harming the environment and that something must be done to improve waste management practices. Problem recognition is often marked by a catalyzing event, such as the discovery of buried hazardous waste at Love Canal in the United States or the 1989 publication in Hong Kong of the white paper *Pollution in Hong Kong: A Time to Act,* which focused national attention on the problem of hazardous waste. In both cases, the government was spurred to act—in the United States, to enact Superfund legislation and in Hong Kong, to develop comprehensive hazardous waste regulations and infrastructure.

As in the U.S. and Hong Kong cases, legislation requiring the development and implementation of a hazardous waste program is typical of this first stage. The developed countries we examined all established the legal framework for their HWM programs in specific hazardous waste laws. In contrast, all of the developing countries we studied promulgated hazardous waste regulations under the authority of existing, more general, environmental laws (sometimes after amending them with specific provisions concerning hazardous waste). Many of these more general laws were passed a decade before public attention turned to HWM issues. Figure 2-1 shows the date of each country's first major legislative or regulatory effort to manage hazardous waste.

Over time, new laws and amendments add new elements to, or change the emphasis of, HWM programs. In the countries we studied, this new legislation often focused on two issues: cleaning up sites contaminated with hazardous substances and focusing the goals of HWM policies on recycling and pollution prevention. A third issue, that of harmonization, emerged for Germany and Canada, both of which started their HWM programs with significant delegation of authority to subnational jurisdictions. In Germany, for example, the 1986 Waste Avoidance and Management Act harmonized many of the standards that had been set independently by states under the 1972 Waste Disposal Act. More recently, Germany and Denmark have begun the process of harmonizing their national hazardous waste management policies with the other members of the European Union.

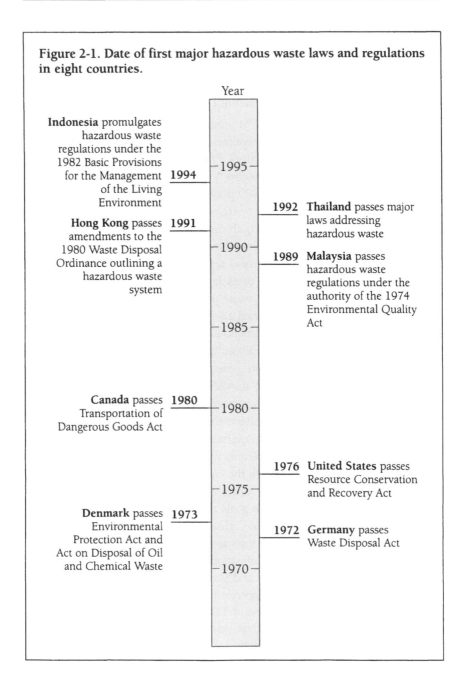

Figure 2-1. Date of first major hazardous waste laws and regulations in eight countries.

Designating a Lead Agency

Legislative mandates for hazardous waste management usually identify an implementing agency and give it regulatory authority. The process of ensuring that an agency can fulfill its mission of carrying out a new program requires addressing at least four issues: (1) creating an agency (or agencies) with the power to regulate; (2) determining how regulatory responsibilities will be shared among national, state, and local authorities; (3) determining whether authority will be concentrated in one agency with environmental responsibilities or distributed across agencies with a variety of responsibilities (for example, agriculture or industry); and (4) ensuring that such agencies have adequate resources—both financial and technical—to carry out their new functions.

A critical element in creating a functioning regulatory system is giving a specific agency the power to regulate. In some countries, like the United States, specific environmental agencies were created with regulatory responsibilities. By contrast, in the developing countries we studied, agencies with regulatory authority typically evolved from existing advisory institutions. These advisory institutions recommended policies and coordinated the actions of ministries with environmental responsibilities, but had no direct authority to enforce environmental laws. In these countries, making the transition from an advisory body to a regulatory agency provided a strong signal about the seriousness of the regulatory program.

Vesting power in an implementing authority also requires determining the allocation of relationships and responsibilities between national and subnational authorities (for example, states and local governments). In most of the countries we studied, federal authorities had the responsibility for designing programs, while implementation and enforcement were delegated to state or local authorities. In Denmark, for example, much of the responsibility for inspecting and sanctioning facilities is vested in local municipalities, while the national authority sets standards and settles disputes between industries and local regulators. In the United States, the federal government sets national standards for hazardous waste management, but the U.S. Environmental Protection Agency (EPA) has authorized most states to implement their own HWM programs (in lieu of the federal program) after EPA certifies that these programs meet or exceed federal requirements. In Canada and Germany, devolution to provincial and state authorities is much more extensive, giving them most of the responsibility for designing and implementing HWM programs in their jurisdictions.

Each country also must determine whether to vest implementation and regulatory responsibilities in one agency or in a number of agencies. All but one of the countries we studied gave authority to one environmental regulatory body. The exception is Thailand, where regulatory and enforcement responsibility is shared among a number of government bodies and jurisdictions, each obtaining its authority from different laws concerning hazardous waste management. This diffusion of responsibility and lack of coordination are regarded by a leading environmental research organization in Thailand as major weaknesses of the country's hazardous waste regulatory system.[1]

Finally, any regulatory agency needs the resources and trained staff to carry out its mission. In developing countries, this is a particularly important issue. Indonesia's regulatory agency reportedly has had difficulty attracting well-trained and experienced staff.[2] In Malaysia, environmental authorities have had difficulty keeping trained personnel from leaving for more lucrative consulting positions in the private sector.[3] In Thailand, a shortage of trained personnel, technology, and know-how has been noted as a significant constraint on proper waste management.[4] In fact, a regulatory official in Thailand stated in 1996 that the country had insufficient experienced personnel to run the four hazardous waste treatment and disposal facilities planned at that time.[5] Even in developed countries, maintaining high-quality staff can be a challenge. In the United States, EPA and state agencies often serve as training grounds for HWM professionals. Once staff are experienced in these issues, they are often recruited by private industry, at much higher salaries.

Promulgating Rules and Regulations

Once a hazardous waste law has been enacted and a lead agency has been designated, regulations and requirements must be developed to make the new program a reality. Often, requirements are phased in over time as the regulatory program develops. Typically, these rules and regulations have a number of common elements, each of which is discussed below. Because the *definition* of hazardous waste is such an important (and often controversial) aspect of hazardous waste regulation, we discuss it in some depth first. We then go on to outline the many aspects of regulating the *management* of hazardous waste from generation to storage, transportation, treatment, and disposal.

Defining Hazardous Waste

The first step in any hazardous waste program is defining which wastes will be regulated and what authority, if any, subnational governments will have to

develop their own definitions and exemptions. There is no clear-cut or generally agreed-on definition of hazardous waste. Even the names are different:
"hazardous waste" in the United States and Thailand, "special waste" in
Germany, "scheduled waste" in Malaysia, "dangerous goods" in Canada, "B3
waste" in Indonesia, "chemical waste" in Hong Kong, and "waste chemicals" in
Denmark.

Countries use a variety of approaches to classify wastes as hazardous.
These approaches include classification by chemical characteristics (such as
ignitability or explosivity), categories of industrial by-products that have undergone specific processes (such as fly ash), technology of origin (such as electroplating), generic groupings (such as oily waste), and specific lists of wastes or
waste types.[6] Most countries also have specific lists of wastes that are exempt
from hazardous waste regulation. Different classification schemes and exemptions lead to widely varying lists of regulated wastes among countries.

Because hazardous waste definitions (and the quality of data reporting on
generation) vary widely among countries, making comparisons among countries
is notoriously difficult. For example, some countries (such as the United States)
include hazardous wastewater in their definition of hazardous waste, while others do not. This makes a tremendous difference in the total volume of reported
hazardous waste. Total hazardous waste generated in the United States in 1995,
for example, was 279 million tons, of which all but 12 million tons was wastewater.[7]

Figure 2-2 shows each country's hazardous waste generation as a function
of gross national product.[8] The volume of hazardous waste used in the figure was
determined according to each country's own definition. The numbers in Figure
2-2 may tell us more about the differences in how hazardous waste is defined in
each country than the comparative magnitude of each country's hazardous waste
problem. The data can, however, be used to compare the size of the HWM effort
each country has laid out for itself relative to its economic output.

When subnational authorities, or even different agencies within a country,
have authority to define hazardous wastes, the definition of what is regulated can
vary widely even within a country. In the United States, for example, several
states have their own definitions of hazardous wastes that are more inclusive
than the federal definition. Historically, Canadian provinces and German states
have had significant discretion in determining which wastes to regulate as hazardous. In Thailand, the environmental regulatory authority uses a different definition of hazardous waste than the Ministry of Industry, which has significant regulatory authority over industry. Because different definitions complicate the reg-

Figure 2-2. Hazardous waste generation and gross national product (GNP).

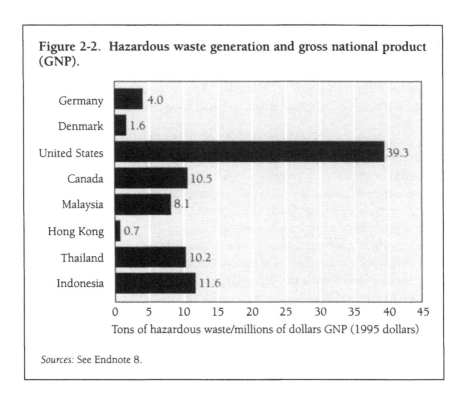

Tons of hazardous waste/millions of dollars GNP (1995 dollars)

Sources: See Endnote 8.

ulatory system—particularly when waste is transported across jurisdictions with different definitions—recent trends have been toward the national harmonization of waste definitions. National efforts at harmonization were undertaken in Germany in the mid-1980s, and a significant effort at harmonization has been undertaken in Canada in the last few years. European Community directives have helped harmonize waste lists in Europe, while the Basel Convention and other international efforts have helped establish consistency throughout the world.

Regulating Hazardous Waste

Beyond defining hazardous waste, regulations must address a number of other aspects of how waste is managed, such as:

- *Requirements for hazardous waste generators.* Requirements for generators generally outline technical and/or performance standards for proper storage, treatment, and disposal of wastes, as well as systems for registering

generators with a regulatory authority. They also identify which generators will be exempted from these regulations and how data on waste generation will be reported to the government.

- *Requirements for hazardous waste treatment, storage, and disposal facilities.*
  These requirements outline standards for treatment, storage, and disposal facilities and establish procedures for government licensing and permitting. There usually are also record-keeping and reporting requirements for HWM facilities.
- *Requirements for transporters of hazardous waste.* These requirements typically involve rules for licensing transporters and placarding vehicles.
- *Manifests for all off-site shipments of waste.* These requirements outline
  manifest systems by which all transfers of hazardous wastes from generators to transporters to treatment, storage, and disposal facilities are
  recorded and tracked.
- *Fines, penalties, and enforcement actions.* These regulations outline the systems for enforcing regulatory requirements, including inspections, specific types of enforcement activities, and fines or other penalties for violations of various aspects of the regulatory system.

Promulgating these types of regulations can take some time. In the United
States, it took three to six years to develop the Resource Conservation and
Recovery Act's (RCRA's) core regulations. In Indonesia, it took seven to eight
years, and in Malaysia, it took five years. Once regulations are in place, they
often require periodic modification, as occurred in the United States throughout
the 1980s and continues today. Often, more stringent requirements are phased
in over time.

Establishing the institutions and programs that support the regulatory system also takes time. It requires personnel, training, bureaucratic support, and
adequate resources. In Germany, for example, it took at least three years for state-
level officials to become fully familiar with the operation of the manifest system
they were supposed to administer.[9]

In a number of countries, regulations were promulgated that required better waste management practices, but government chose to let enforcement lag
behind regulation to give generators of hazardous waste and operators of waste
management facilities time to comply with the new, more stringent requirements. In the United States, this is referred to as "enforcement discretion." This
approach works if the threat of future enforcement is perceived to be real.

Developing Treatment and Disposal Capacity

Although most of the stages of a regulatory program discussed thus far have been roughly sequential, the development of treatment and disposal capacity can occur at any stage in the process. The dynamic between the evolution of a regulatory program and the development of treatment and disposal capacity is complex. In fact, it is a typical chicken-and-egg question. Without stringent regulations and enforcement, there is little incentive to develop (or use) state-of-the-art hazardous waste treatment and disposal facilities—but, without such facilities, it is hard to comply with stringent waste management requirements.

When a regulatory program precedes infrastructure, countries must take interim measures—such as storage, codisposal, or export—until adequate capacity has been developed. Countries do not necessarily need to develop their own hazardous waste treatment and disposal infrastructure; much hazardous waste crosses national borders to be managed elsewhere. For example, approximately 600,000 metric tons of hazardous waste are shipped annually across the U.S.–Canadian border for treatment and disposal in permitted facilities in the two countries.* Around 13% of the hazardous waste received at Denmark's Kommunekemi facility in 1995 was imported from Germany, the Netherlands, and Norway.[10] Still, while the transboundary movement of hazardous waste is a significant component of many countries' HWM systems, it is generally not until at least some domestic capacity exists that countries' regulatory programs can be fully implemented.

When infrastructure precedes an effective regulatory system, often there is not adequate demand for the facility to operate at a profit, much less break even. It can take years before generators are willing (or forced) to pay the full cost of treatment and disposal. Faced with this prospect, many private investors are unlikely to finance the construction of hazardous waste facilities. This is especially true in developing countries with nascent judicial and enforcement systems that create considerable uncertainty about whether regulatory programs will be enforced.

In the countries we studied, there were examples of infrastructure preceding regulation, of simultaneous development of regulations and infrastruc-

---

*Canada imports approximately 400,000 tons of hazardous waste per year from the United States, and the United States imports approximately 200,000 tons per year from Canada (communication between Dave Campbell [Environment Canada] and K. Probst, January 5, 1999).

ture, and of infrastructure lagging behind regulation. Because the timing of regulation and infrastructure is so important, we go into some detail on the various approaches taken by the countries in our study.

The cases of the Mittelfranken Cooperative in Germany's state of Bavaria and the Stablex facility in Quebec illustrate the situation of infrastructure preceding regulation. Bavaria's Mittelfranken Cooperative for Special Waste Management was established in 1966 and developed a centralized treatment, storage, and disposal (TSD) facility for the highly industrialized Mittelfranken district in 1968—four years before Germany's 1972 Waste Disposal Act and before the establishment of the national environmental agency. The effort, led by local government, was spurred largely by local concerns about water pollution from the uncontrolled disposal of hazardous waste.[11] The experience gained in Bavaria influenced the development of national hazardous waste policy significantly. By the late 1980s, the Bavarian system was widely regarded as being effective.[12]

One risk of building facilities before the development of a regulatory system is lack of demand for the facility. This was the case in Quebec where a private firm, Stablex, chose to build a TSD facility near Montreal with the expectation that the province would soon be implementing a hazardous waste regulatory system.[13] The facility was built in 1983, but the promulgation of laws was not complete until 1985. The timing appears to have affected the financial performance of the facility; it was not until 1987 that it turned a profit.[14]

In a sense, some of the United States' hazardous waste infrastructure existed before regulation, but there is a very important difference between the U.S. case and the cases of Bavaria and Quebec. In the United States, a number of existing facilities were converted to HWM facilities after RCRA regulations took effect. These facilities included cement kilns, which provide much of the U.S. capacity for incineration of hazardous waste, and landfills, which were upgraded over time to meet permitting requirements for the land disposal of hazardous waste. Many of the poorest environmental performers among these landfills shut down in the early 1980s rather than comply with the more stringent requirements phased in under RCRA.

Hong Kong and Indonesia illustrate the case of governments' coordinating the promulgation of regulations with the development of facilities. Hong Kong phased in hazardous waste regulations in 1993 at the same time that it opened its first modern hazardous waste treatment and disposal facility, the Chemical Waste Treatment Center. In Indonesia, the promulgation of hazardous waste regulations in 1994 and 1995 coincided with the opening of the PT Prasadah Pemunahan Limbah Industry (PPLI) integrated HWM facility near Jakarta.

While the Hong Kong facility was an early success, the Indonesian facility has been disappointing. As we discuss in Chapter 3, a major cause of this difference in performance was the fact that disposal in Hong Kong was free (due to government subsidies), while in Indonesia, it was relatively expensive by local standards. It is also quite likely that generators in Hong Kong felt the credible threat of enforcement more strongly than generators in Indonesia. The difference between the two countries' experiences highlights another important point: Even when the timing of regulation and facility construction is coordinated, there is still likely to be a period of uncertain demand until procedures for identifying facilities and enforcing regulatory requirements are instituted. To some extent, Hong Kong prepared itself for this compliance lag by instituting a set of "interim arrangements" in the early 1980s that provided information on waste generators and allowed regulatory personnel to gain experience. When Hong Kong's TSD facility opened, regulators already had information on generation and some experience with hazardous waste management.[15]

The final model is one in which regulation preceded the development of facilities. In the countries we examined, the best illustration of this is Malaysia. Malaysia began the process of seeking a private company to construct and operate a hazardous waste facility following a feasibility study on hazardous waste management in 1987. However, the country's first HWM facility (run by the firm Kualiti Alam) has only recently opened, eleven years later. In the interim, the country promulgated and began to enforce hazardous waste regulations, starting in 1989. With regulations in place, but no modern facility for the proper treatment and disposal of waste, Malaysia's regulatory authority had to make arrangements for storing hazardous waste on site, for minimizing generation, for shipping waste abroad, or, in the case of a few large facilities, for incinerating waste on site.[16] Since the early 1990s, Malaysia has also licensed a number of smaller, "environmentally sound," off-site secure landfills and off-site pretreatment, recycling, and recovery facilities (159 were licensed in 1997 alone).[17] Illegal dumping, however, is thought to persist, particularly among small generators.[18]

Creating a Mature Compliance and Enforcement Program

In the final stage of a regulatory program, all elements of the system are fully operational. Laws exist, regulations have been developed and are being enforced, generators have been identified, and treatment and disposal facilities have been constructed. Something more than a formal regulatory program is needed, however, to create an effective regulatory program: a country needs a "culture of compliance." When such a culture exists, compliance is the norm; that is, the majority of firms handle, store, transport, treat, and dispose of hazardous waste properly.

No government has enough resources to inspect every HWM facility frequently enough to detect all violations of rules and regulations. Instead, the regulatory system must have enough public credibility—often established by making the threat of enforcement real—that most regulated entities comply with environmental requirements as a matter of course. Although, in theory, it might be possible to achieve a culture of compliance for environmental programs only, in practice, the success or failure of environmental programs depends on the larger political and legal culture of a country.

In some countries, the rule of law prevails, while in others it is still weak. In countries where bribery and collusion between the regulators and the regulated are commonplace, it is difficult to imagine how many of the features of a hazardous waste—or any other—regulatory system can function well. Some countries have experimented with ways to compel compliance in the absence of an effective legal system, such as using public information to generate public pressure on firms to control pollution. Preliminary results from such projects are encouraging, but it is clear that a weak legal system, the lack of enforcement, or a culture that tolerates corruption makes the task of developing an effective HWM system more difficult.

All of the developed countries we studied can be said to have achieved a culture of compliance by the end of the 1980s. Most analysts agree that the United States reached this point in the late 1980s. The timing is similar for Denmark, where its system of supervision and inspection became largely operational in 1987.[19] In Germany, the date is sometime after 1986, when the Waste Avoidance and Management Act substantially revised the 1972 Waste Disposal Act—mainly because the earlier law was not leading to effective hazardous waste management.[20] Environmental officials in Quebec and Alberta have said that their provincial HWM programs were also largely in place by the end of the 1980s.[21] However, even in these countries, where most would acknowledge that the regulatory systems are effective, it is difficult to tell specifically how well the system is working. Illegal dumping is a continuing problem in the United States, mostly among small-scale operators.[22] In Canada, as late as 1993, some analysts believed that only a small volume of reported waste volumes was actually being managed appropriately.[23]

None of the developing countries we examined have reached the point where their regulatory systems would be regarded as fully operational. Hong Kong appears to be the furthest along. Procedures for registering generators, licensing transport and treatment and disposal facilities, and conducting inspections appear to be well developed. To date, Hong Kong has registered 9,200 chemical waste producers, and officials believe that this covers most generators. Additionally, officials have licensed seventy chemical waste collectors, and thirty-four chemical waste treatment and disposal facilities.[24] In 1996, the environmental agency conducted over 7,000 inspections.[25]

Although Indonesia, Thailand, and Malaysia have succeeded in starting HWM programs, they appear to be a number of years away from addressing the majority of their respective hazardous waste problems and changing the behavior of the majority of generators. There are, however, some promising signs in each country. Indonesia, for example, has recently increased the enforcement authority of its environmental regulatory agency, BAPEDAL. In Malaysia, regulators brought 275 cases against violators of its Environmental Quality Act in 1997. However, significant problems remain in all three countries: the current hazardous waste treatment and disposal capacity is inadequate to handle the amount of hazardous waste generated; regulatory agencies lack resources and trained personnel; and many firms generating hazardous waste balk at paying disposal fees. Until these problems are addressed, a mature compliance and enforcement system will remain an elusive goal.

Patterns of Program Development in Developed and Developing Countries

Some general patterns emerge from the experiences of the eight countries we examined. Denmark, Germany, and the United States passed their first major hazardous waste laws between 1972 and 1976, with Canada following in 1980. A five- to seven-year period followed during which these countries developed hazardous waste regulations and requirements. In the early to mid-1980s, the United States, Denmark, and Germany refined and expanded their regulatory programs through another round of hazardous waste laws or major amendments to existing laws. In Canada, the federal government, Alberta, and Quebec passed their hazardous waste regulations around the same time.

By the end of the 1980s, the regulatory systems in all four developed countries were mostly operational. An optimistic assessment of the time it took these countries to develop mature programs is from ten to fifteen years. Subsequent laws and policies developed during the 1990s focused mainly on waste minimization and recycling, as well as harmonization with international standards and the cleanup of contaminated sites.

The evolution of regulatory programs in Hong Kong, Indonesia, Malaysia, and Thailand followed a similar, but more recent, pattern. By the early 1980s, all of these countries had enacted some form of environmental legislation with at least some authority to regulate hazardous waste. However, hazardous waste management received little attention until the late 1980s, after periods of rapid economic growth and the expansion of the countries' manufacturing sectors. From 1989 to 1995, all of the developing countries we examined passed major new legislation addressing hazardous waste or developed regulations outlining comprehensive programs: Malaysia in 1989, Hong Kong in 1991, Thailand in 1992, and Indonesia in 1994 and 1995.

If the wave of regulations and laws in the late 1980s and early 1990s is taken as the starting point of program development in these four countries, the process has, thus far, taken from five to ten years. However, all of these countries actually began their efforts to develop a hazardous waste regulatory system some years before regulations were finally promulgated. In Malaysia, for example, an early draft of hazardous waste regulations was written in 1984. In Indonesia, a first draft of hazardous waste regulations was written in 1987. If these earlier dates are used to mark the starting point of program development, these countries are already reaching the ten- to fifteen-year time horizon experienced by the developed countries.

With the possible exception of Hong Kong, the developing countries we studied are still some way from having what most would describe as fully functional programs. The number of additional years needed until this stage is reached is, of course, unknown.

Lessons Learned

Given the differences among the eight countries we studied and the varied approaches taken by each to establish HWM programs, it is perhaps surprising that general lessons can be derived from an examination of program development. Perhaps the most significant lesson is one of humility about the enormity of undertaking the establishment of a hazardous waste management program. We turn to this issue first.

1. It takes a long time—at least ten to fifteen years—to develop a fully operational hazardous waste regulatory system.

It took the United States, Germany, Canada, and Denmark ten to fifteen years to develop the laws, institutions, and procedures that resulted in widespread changes in the way firms handle, treat, and dispose of hazardous waste. Because it is so difficult to identify the end point on this scale, the time period may have been even longer. These four are "rule-of-law" countries in which a strong culture of compliance and strong public support for the goals of the regulatory program exist. Although ten to fifteen years sounds like a long time, in most countries, enacting legislation and promulgating implementing regulations are each multiyear processes. When a few years are added to actually start implementing new requirements and another few years to revise them based on implementation experiences—and a few more to build and permit needed HWM facilities—ten to fifteen years is not, in fact, such a long time.

The developing countries' experiences confirm that even with the experience gained from developed countries, it still takes a long time to develop an effective HWM program. Although Hong Kong, Indonesia, Malaysia, and Thailand have been actively regulating hazardous waste for only five to ten years, most began their efforts to write regulations or to find appropriate locations for facilities in the early 1980s. These four countries, with the possible exception of Hong Kong, still have a long way to go before hazardous waste is properly managed in a comprehensive way.

2. Developing a culture of compliance is the crucial element of an effective hazardous waste management system.

Absent a culture of compliance, it is extremely difficult to reach the final stage of an HWM program: a mature compliance and enforcement program in which compliance is the norm and the behavior of waste generators has changed from an unregulated environment. There is little incentive for generators to pay any increase in cost for proper waste management if there is not some expectation that they will be held accountable for meeting new, more stringent, standards. The question of what, if anything, can be done to create a culture of compliance is a very difficult one and is much broader in scope than the challenge of creating a hazardous waste regulatory program. At its core, a culture of compliance is the product of how government functions and whether government is perceived as credible—areas warranting further research.

3. Clear lines of regulatory authority increase the chances of successful implementation of new regulatory programs.

A single *environmental* agency may not always be the best way to proceed—particularly if a new agency can be expected to have little power in the bureaucracy—but having clear lines of regulatory authority vested in a single agency is important to program success. The experience of Thailand, with its multiple regulatory bodies, is sobering. To some extent, the lack of clear lines of authority—as well as inadequate funding and a lack of trained personnel—characterizes Hong Kong, Malaysia, and Indonesia, as well. Laws and regulations are not enough for programs to be successful. Institutions must also be strong, and agencies need the authority and the resources to effectively regulate and enforce.

4. There are important consequences to the decision not to harmonize policy at the national level.

Most of the countries we studied had some form of shared responsibility between the national government and subnational jurisdictions. Two countries—Canada and Germany—delegated most of the responsibility for the design and implementation of programs to provinces and states, reflecting the traditional allocation of government responsibilities in these two nations. In both countries, decentralization led to large differences in the quality and effectiveness of different state- and province-level programs, as well as to differences in waste definitions, permitting requirements, and treatment and disposal standards.

Other countries have also experienced some frustration with decentralized HWM functions. In the United States, for example, data reporting, which is generally the responsibility of states, has been problematic. Because of states' differing definitions of waste, incomplete lists of facilities, and the omission of data, there is little comprehensive information on long-term nationwide trends in hazardous waste generation, management, and disposal.[26] In Denmark, one of the consequences of decentralization reportedly has been a lack of uniformity of enforcement by the various municipalities.[27]

In general, the question of decentralization raises important issues of institutional capacity. Assigning regulatory authority to state or local governments requires that these levels of government have the financial and human resources to develop effective institutions and carry out programs. As mentioned earlier, it has been difficult for many of the developing countries we studied to marshal these resources at the national level, much less across a number of local jurisdictions.

Endnotes

1. Thailand Development Research Institute (TDRI), *The Monitoring and Control of Industrial Hazardous Waste: Hazardous Waste Management in Thailand* (Bangkok: TDRI, 1995), p. 12.

2. Environmental Resources Management (ERM), *Public/Private Sector Cooperation in the Provision of Hazardous Waste Management Facilities* (London: ERM, 1994), p. A2.

3. Communication between Gordon Young (United States–Asia Environmental Partnership, Malaysia) and T. Beierle, May 20, 1998.

4. International Maritime Organization (IMO), "National Waste Management Profile for Thailand," in *Global Waste Survey: Final Report* (London: IMO, 1995), p. 183.

5. "Chemcontrol Symposium II: Developing Markets," *Haznews,* No. 104 (November 1996).

6. Ray Kopp, Paul Portney, and Diane Dewitt, "International Comparisons of Environmental Regulation," in *Environmental Policy and the Cost of Capital* (Washington, D.C.: American Council for Capital Formation, 1990), pp. 80–81.

7. U.S. Environmental Protection Agency (EPA), "Treat, Store, and Dispose of Waste" (1998) (available at http://www.epa.gov/epaoswer/osw/tsd.htm).

8. Gross national product data are from World Bank, *World Development Report 1997* (Washington, D.C.: World Bank, 1997). Hazardous waste data are from the following sources:

Germany (1993 data): Institute for Prospective Technological Studies (IPTS), *The Legal Definition of Waste and Its Impact on Waste Management in Europe* (Seville, Spain: European Commmission–Joint Research Center, IPTS, 1997), p. 21.

Denmark (1995 data): Ibid.

United States (1995–1996 data): EPA, "Treat, Store, and Dispose of Waste."

Canada: Environment Canada, *Status Report on Hazardous Waste Management Facilities in Canada—1996* (Ottawa, Ontario: National Office of Pollution Prevention, February 1998), p. 8.

Malaysia (1996 data): E-mail from Lim Thian Leong (Center for Environmental Technologies, Malaysia) to T. Beierle, May 8, 1998, referring to data collected by Department of Environment.

Hong Kong: E-mail from R.C. Rootham (Local Control Office [Territory East], Environmental Protection Department, Hong Kong) to K. Probst, December 5, 1998.

Thailand (1996 data): E-mail from Sombat Sae-Hae (Thailand Development Research Institute) to T. Beierle, May 27, 1998, reporting data from Division of Industrial Hazardous Waste Management.

Indonesia: United States–Asia Environmental Partnership, "US-AEP Country Assessment: Indonesia" (available at http://www.usaep.org/country/indonesia.html).

9. F. Van Veen, "National Monitoring Systems for Hazardous Waste," in *Transfrontier Movements of Hazardous Waste* (Paris: Organisation for Economic Co-operation and Development, 1985), p. 84.

10. "Profits down 48% for Kommunekemi," *Haznews*, No. 89 (August 1995).

11. Bruce Piasecki and Gary A. Davis, "A Grand Tour of Europe's Hazardous Waste Facilities," *Technology Review,* Vol. 87 (July 1984).

12. Joanne Linnerooth and Allen V. Kneese, "Hazardous Waste Management: A West German Approach," *Resources* (Summer 1989).

13. R.J. Cooke, "Canada's Hazardous Waste Infrastructure," *EI Digest* (November 1993), pp. 3-4.

14. Barry G. Rabe, *Beyond NIMBY: Hazardous Waste Siting in Canada and the United States* (Washington, D.C.: Brookings Institution, 1994), p. 125.

15. ERM, *Public/Private Sector Cooperation*, p. C7.

16. Ibid., p. B1.

17. E-mail from Lim Thian Leong to K. Probst, December 9, 1998.

18. ERM, *Public/Private Sector Cooperation,* pp. B1 and B4.

19. Mogens Moe, "Environmental Administration in Denmark," *Environment News,* No. 17 (1995), Section.13.7.2 (available at http://www.mem.dk/mst/books/moe/).

20. Joachim Wuttke, "Management of Hazardous Waste in an Industrialized Society," 2nd International Conference on Environmental and Industrial Toxicology: Research and Its Application, Bangkok, Thailand, December 9–13, 1996, p. 2.

21. For Quebec, E-mail from Benoit Nadeau (Ministère de l'Environnement et de la Faune, Quebec) to T. Beierle, June 10, 1998; for Alberta, communication between Antonio Fernandes (Alberta Environmental Protection) and T. Beierle, May 27, 1998.

22. Bette Hileman, "The Downside of Environmental Rules: Opportunities for Crime," *Chemical and Engineering News,* Vol. 76, No. 38 (September 21, 1998), p. 41.

23. Cooke, "Canada's Hazardous Waste Infrastructure," pp. 5–7.

24. E-mail from R.C. Rootham to T. Beierle, May 11, 1998; E-mail from K.K. Yung (Environmental Protection Department, Hong Kong) to T. Beierle, April 25, 1998.

25. Environmental Protection Department, *Environment Hong Kong 1997* (Hong Kong: Environmental Protection Department, 1997), p. 222.

26. J. Clarence Davies and Jan Mazurek, *Pollution Control in the United States: Evaluating the System* (Washington, D.C.: Resources for the Future, 1998), pp. 79–80.

27. Moe, "Environmental Administration in Denmark."

3. Facility Financing

As previously noted, one of the major challenges to getting a hazardous waste management (HWM) program up and running is ensuring adequate waste treatment and disposal capacity so that generators can, in fact, comply with regulations requiring sound waste management. While there is often general agreement that new and better hazardous waste treatment, storage, and disposal (TSD) facilities are needed, there are three fundamentally different approaches to how these facilities are financed: public-sector financing, private-sector financing, or some kind of public–private joint venture. Two of the key questions we set out to answer are:

1. Which approaches to financing HWM infrastructure have been used by the countries in our study?
2. What are the advantages and disadvantages of the different approaches?

Issues of facility financing relate to the overall performance of the HWM system in two main ways. The first is getting HWM facilities permitted and built. As was noted in Chapter 1, the HWM market is complicated, can take a long time to develop, and is subject to a number of sources of uncertainty. All of these factors discourage private investment. At the same time, treatment and disposal capacity is necessary to get regulatory programs up and running. Several countries have responded to this tension between an uncertain market and a public policy need by offering a range of incentives to encourage private investment in HWM facilities or by directly investing public money in facility construction and operation.

The second way in which issues of facility financing relate to the overall performance of the HWM system is the role that public-sector investment plays in determining how much generators will need to pay for disposal. By using government subsidies to pay for all or part of facility construction and operation, public investment can reduce the amount that generators must pay to use a facility. This subsidy provides an incentive (the proverbial carrot) to encourage the use of more environmentally sound waste management facilities. The issue of subsidization relates directly to waste management goals because high disposal fees can encourage illegal dumping, especially if regulatory enforcement is weak. Subsidized disposal fees reduce generators' incentives to dispose of waste illegally and can be targeted to sectors that may have fewer resources to pay for disposal—such as small businesses. On the other hand, lowering fees may undermine another important waste management goal, that of encouraging waste reduction.

The following section discusses how issues of facility ownership and operation, incentives, and subsidies define the roles of the public and private sectors in each of the eight countries studied. It is followed by a discussion of how different financing approaches have led to different results in terms of the effectiveness of the HWM systems. The chapter concludes with several lessons learned about facility financing.

Roles of the Public and Private Sectors in Financing Hazardous Waste Management Facilities

Some countries view hazardous waste management as a private good to be provided in the marketplace, while others see it more as a public good to be provided as a government service. In the private-sector approach, companies pay for hazardous waste management much as they would pay for any other service, such as building construction, equipment maintenance, or the resurfacing of a parking lot. Private firms construct HWM facilities and charge disposal fees that cover capital and operating costs. In the public-service approach, government provides HWM services using tax revenues, just as many governments provide solid waste management, wastewater treatment, or street-cleaning services.* Taxes to pay for the service may target industries most likely to generate hazardous waste, through a tax on chemical feedstocks, for example. Alternatively, funds may come from general taxes, justified by the argument that sound hazardous waste management provides widespread public benefits by keeping waste out of open dumps, surface water, and groundwater.

There are a number of reasons why countries might differ in thinking of HWM as a private or public good. These include the government's ability (and willingness) to institute and collect taxes and the financial resources of generators. Perhaps most important is each country's political ideology and the extent to which government has traditionally played a role in industrial policy. For example, the United States does not generally subsidize private firms directly, whereas such subsidies—or, often, joint public–private partnerships—are more common in Europe.

The countries we studied ran the gamut from a purely private-sector to a purely public-sector approach, but most fell somewhere between the private and public approaches. In some cases, private waste management firms received free land or other incentives from government. In other cases, public companies charged full (unsubsidized) disposal fees as if they were private companies. Various arrangements of ownership and operation, the provision of investment incentives, and the question of subsidies define an often-complicated set of shared responsibilities between the public and private sectors. Each of these aspects of the public–private relationship is discussed below.

Facility Ownership and Operation

Hazardous waste treatment and disposal facilities are owned and operated by the private sector, the public sector, or some combination of the two.[1] The United States, Indonesia, Malaysia, and Canada's Province of Quebec all have relied largely on the private sector for the ownership and operation of HWM facilities (see chart, "Type of Facility Ownership and Operation by Country"). This private-sector approach has a number of potential advantages, most centered on the potential efficiency and expertise of the private sector compared

---

*It should be noted that these are services typically provided by local, rather than national, governments and that all residents use these types of services directly, which is not the case for HWM facilities.

Type of Facility Ownership and Operation by Country[a]

Private Ownership and Operation
- United States (all facilities)
- Quebec, Canada (Stablex)
- Indonesia (PPLI)[b]
- Malaysia (Kualiti Alam)

Mixture of Public and Private Ownership and Operation

Public–private ownership and private operation
- Thailand (Rayong)
- Alberta, Canada (Swan Hills)[c]

Public–private ownership and operation
- Hesse, Germany (HIM)
- Bavaria, Germany (GSB)

Public ownership and private operation
- Hong Kong (CWTC)

Public ownership and public–private operation
- Thailand (Samae Dam)

Public Ownership and Operation
- Denmark (Kommunekemi)
- Bavaria, Germany (ZVSMM)

---

[a] Chart reflects early years of operation of these facilities and only includes facilities discussed in this report.

[b] Indonesia's PPLI facility is mostly private: it has a 5% ownership stake by the country's regulatory agency, BAPEDAL.

[c] Alberta privatized the Swan Hills facility in 1996.

Notes: CWTC, Chemical Waste Treatment Center; GSB, Gesellschaft zur Entsorgung von Sondermull; HIM, Hessische Industriemull GmbH; PPLI, PT Prasadah Pemunahan Limbah Industry; ZVSMM, Mittelfranken Cooperative for Special Waste Management.

with the public sector. With fewer bureaucratic entanglements, private firms may be able to build facilities more quickly and operate them more efficiently. Private companies are likely to have more experience than government agencies in developing and operating similar facilities elsewhere. The private-sector approach also places the responsibility for raising capital and absorbing financial risk on private companies rather than on taxpayers. Because they bear potential losses more directly than taxpayers, these private investors may be expected to exercise greater diligence in making appropriate investment choices to ensure the profitability of facilities.

However, the private-sector approach has several potential disadvantages. Private-sector firms need to make money. How much HWM capacity the private sector will build depends on the financial returns of the investments, *not* on the urgency of solving a country's hazardous waste problem. The willingness to invest, in turn, depends on generators' willingness (and ability) to pay the full cost of disposal, because disposal fees generally are not subsidized in the private-sector approach. In countries with new HWM programs, companies may be unwilling to invest without some commitment from government to share financial risk or offer other incentives (we return to the topic of incentives later in the chapter). Moreover, some types of wastes are so controversial or expensive to treat—dioxin wastes or wastes containing radioactive material are examples in the United States—that a private market is simply unlikely to develop. Private-sector owners also may be subject to greater risk of financial failure. If firms go bankrupt or choose to close a facility, a country could be left with no treatment and disposal services.

On the other end of the public–private spectrum are publicly owned and operated facilities, such as Denmark's Kommunekemi facility and Bavaria's Mittelfranken Cooperative (ZVSMM).† The advantages and disadvantages of the purely public-sector approach are largely the reverse of those of the private-sector approach. The ability to use government funds and to obtain low rates on capital (from multilateral lending institutions in the case of developing countries) makes the viability of public companies more assured. Public funding allows governments to bring needed facilities on-line in the early stages of regulatory programs, when the regulatory regime is still weak and a market for waste treatment and disposal is still maturing. Because they have access to government

---

†The Mittelfranken Cooperative is now the Sonderabfallentsorgung Franken mbH (SEF) (fax from Dr. Joachim Wuttke [Umweltbundesamt] to K. Probst, June 17, 1998).

funds, publicly funded facilities have the option of paying for some or all of a facility's capital and operating costs. This gives them more flexibility in determining disposal fees. Public subsidies allow policymakers to shift disposal costs to different parts of the economy to pursue public policy goals (for example, subsidizing small businesses with tax revenues). Less expensive (subsidized) disposal helps to encourage better waste management practices on the part of generators.

Most of the disadvantages of the public-sector approach relate to the potential inefficiencies commonly associated with bureaucracies. Government agencies are typically slower to act than the private sector, in large part because of the political difficulty attached to controversial decisions. In addition, large bureaucracies can be unwieldy and subject to various constraints unique to government (for example, hiring, firing, and contracting rules), which can make it difficult to take the actions needed expeditiously, whether this be hiring staff with needed expertise or selecting a contractor to build a facility. Moreover, if the public entity regulating a facility is not sufficiently independent of the organization operating the facility, the facility's environmental performance may be compromised. Finally, a lack of transparency in how public money is being spent may encourage corruption.

The third approach involves both the public and private sectors. Various combinations of private companies, public entities, and public–private joint ventures have been used to build and operate HWM facilities in Thailand; Alberta, Canada; Hesse, Germany; one of the waste management firms in Bavaria, Germany (Gesellschaft zur Entsorgung von Sondermull, or GSB); and Hong Kong. The main benefit of these mixtures of public and private roles is the ability to retain some of the advantages of a public approach—specifically, the stability that comes with lower-cost financing and the ability to support the facility using government funds—while still benefiting from the advantages of a private-sector approach—specifically, the efficiency and expertise of private firms.

Alberta and Hong Kong are good examples of this combined approach. In Alberta, a provincially owned "crown corporation" partnered with a private firm in a joint venture to construct and operate the province's Swan Hills TSD facility. While the private firm managed the day-to-day operations of the facility, the crown corporation provided the vehicle for transferring public funds to the joint venture for construction and operation. Like Alberta, Hong Kong used a combination of public and private involvement to construct its main HWM facility. The government contracted with a private firm to build and operate the facility. While the private firm paid initial construction costs, the government bought and took ownership of the facility on completion. The private firm was given a fifteen-year operating contract and an arrangement by which the government would reimburse it for operating costs.

In both the Alberta and Hong Kong cases, public entities were able to provide the funding to ensure that facilities were built (and remained in operation), while leaving to private firms the responsibility for day-to-day operation, and—in the Hong Kong case, at least—the responsibility for constructing a facility that met government expectations. The ability to combine the benefits of both public- and private-sector approaches probably explains why public–private joint ventures are the most common approach taken around the world for developing hazardous waste infrastructure.[2] Like the purely public-sector approach, a public–private venture runs the risk of cost overruns (or other results of a lack of accountability) that may occur when owners and operators are not subject to the discipline of the private market. Specific contractual arrangements can help prevent such outcomes; the Hong Kong government's decision to pay its private contractor for the TSD facility only after it was completed, and certified as acceptable, is an example.

Investment Incentives for the Private Sector

Direct financing is not the only way that the public sector can encourage the development of HWM infrastructure. Governments can offer a range of incentives to attract private investment, particularly in the early years of an HWM program. High costs, the potential for public opposition to facilities, and uncertainty about the extent to which regulatory programs will generate demand for a facility's service make hazardous waste management a risky business. Many of these risks are heightened in developing countries where markets are relatively new and uncertain. Offering incentives lowers the financial risk to private investors—decreasing the probability and consequences of not meeting an expected rate of return on investment.

Some risks, of course, such as political instability or natural disasters, are simply part of doing business. Others can be mitigated through contractual or less formal arrangements with government. Powerful incentives are those that reduce uncertainty about future revenues. These include minimum revenue guarantees or exclusive licenses for certain periods of time. Other incentives include project support, such as when government helps with the siting of a facility, facilitates permitting and approval, or provides land. Incentives can also include direct financial support, such as contributions to capital costs (often through a joint venture) or operating costs. Indirect forms of financial support might include tax breaks on equipment, reduced import duties, or a waiver of capital controls to allow offshore borrowing. Government can help firms overcome market barriers, such as transportation costs, by providing hazardous waste pickup, transport, and transfer services. The extent to which particular incentives or groups of incentives are sufficient to encourage private investment varies from case to case, and they are typically the subject of individual contractual negotiations between HWM firms and government.

Table 3-1 shows the types of incentives provided in the four developing countries we examined. It should not be surprising that so many incentives were provided in Hong Kong, where the public and private sectors shared ownership and operation responsibilities. But even in the cases of Indonesia and Malaysia, where the private sector was expected to take the lead in financing hazardous waste facilities, governments provided various incentives to encourage investment.

Disposal Fee Subsidies

While issues of incentives and public-sector ownership and operation affect whether HWM capacity will be available, the government's role in providing subsidies affects the price of disposal. Disposal fees—that is, the amount that generators are directly charged for disposal at a TSD facility—are a crucial aspect of hazardous waste management.[†] Subsidized disposal fees, by using public money to lower the cost of disposal for generators, can encourage compliance. On the other hand, subsidies can discourage another important waste management goal: minimizing the production of waste. In the absence of an effective regulatory regime, then, the question of subsidies raises a trade-off between encouraging compliance and encouraging waste minimization. Whether countries choose to subsidize disposal fees—essentially whether they choose to use pricing policies to emphasize waste minimization or target illegal disposal—can have important impacts on the effectiveness of waste management programs in the early years, when regulation and enforcement are often weak.

[†]For the purpose of this report, we ignore other direct costs involved in hazardous waste management, such as the cost of transporting wastes to TSD facilities, as well as indirect costs.

Table 3-1. Incentives Provided by Government to Private Sector in Four Developing Countries

	Hong Kong	Thailand (Rayong)	Indonesia	Malaysia
Project Support				
Siting support	X	X	X	X
Provision of land	X	X	X	—
Facilitated permitting and approval	—	—	X	X
Financial Support				
Contribution to capital costs	X	X	—	—
Contribution to operating costs	X	—	—	—
Breaks on taxes/duties/capital controls	—	X	X	X
Operating Support				
Government-provided collection and transfer services	X	—	—	—
Risk Management				
Revenue/volume guarantee	X	X	—	—
Exclusive/assured market	—	X	—	X

Note: X, provided; —, not provided.

Sources: See Endnote 3.

When disposal fees are *not subsidized,* the fees charged cover the full capital and operating costs of the facility. *Partially subsidized* fees shift some of the responsibility for costs onto government (and, by extension, the taxpayers). *Fully subsidized* fees shift the entire financial burden of hazardous waste management onto the government, making disposal free for those using the subsidized facility.

To partially or fully subsidize disposal fees, governments can provide capital subsidies, operating subsidies, or both. *Capital subsidies* use public funds to pay some or all of the costs of constructing facilities (and are not subsequently recovered). Such subsidies may take the form of direct investment through outright public ownership or a public–private joint venture, no-interest loans, deferred-payment periods, or some other mechanism. Interpreted broadly, capital subsidies could include any government support that reduces the time or capital investment needed to build a facility, including many of the incentives discussed earlier (such as provision of land, siting support, facilitated permitting, and so forth). In this discussion, however, we limit the definition of capital subsidies to subsidies of the direct costs of constructing facilities.

Operating subsidies use government funding to pay for some or all of the costs of the day-to-day operation of a facility, and this savings is passed on to the users of the facility. In some cases, capital and operating subsidies are combined, giving governments maximum flexibility in determining disposal fees. Only by combining capital and operating subsidies can governments fully subsidize disposal fees.

Table 3-2 shows how the eight countries we examined used three different approaches to subsidization—no subsidies, capital subsidies, and a combination of capital and operating subsidies—in the early years of their HWM programs. Countries are also categorized in the table by the degree of public and private ownership and operation to show the relationship between facility financing and disposal fee subsidies.

The main pattern that emerges in Table 3-2 is that some public role in the ownership and operation of facilities is often combined with subsidization of dis-

Table 3-2. Disposal Fee Subsidies and Facility Ownership and Operation in Eight Countries

Facility Ownership & Operation	No Subsidies	Capital Subsidies	Capital and Operating Subsidies
Private	United States Quebec, Canada (Stablex) Indonesia (PPLI) Malaysia (Kualiti Alam)		
Public and private	Thailand (Rayong) Thailand (Samae Dam)	Hesse, Germany (HIM)[a] Bavaria, Germany (GSB)[a]	Alberta, Canada (Swan Hills)[b] Hong Kong (CWTC)
Public		Denmark (Kommunekemi)[a] Bavaria, Germany (ZVSMM)[a]	

[a] Table reflects early years of operation when the capital costs of facilities in Denmark and Germany were partially funded by government, and these costs were not recovered through disposal fees.

[b] The Swan Hills facility was privatized in 1996, and fees are not currently subsidized.

Notes: CWTC, Chemical Waste Treatment Center; GSB, Gesellschaft zur Entsorgung von Sondermull; HIM, Hessische Industriemull GmbH; PPLI, PT Prasadah Pemunahan Limbah Industry; ZVSMM, Mittelfranken Cooperative for Special Waste Management.

posal fees, but that a public role does not *guarantee* subsidization. Countries with HWM facilities that have some public ownership may or may not choose to subsidize fees. In Thailand, for example, fees are not subsidized; the government recovers full capital and operating costs through disposal fees like a private firm would. The rest of the countries we studied where the public sector has a role in facility ownership and operation do subsidize fees. Some subsidize capital costs only, while the others—Hong Kong and Alberta, Canada—subsidize capital and operating costs.** None of the countries we examined subsidize operating costs but not capital costs. Not surprisingly, the countries with privately owned and operated HWM facilities do not have subsidized disposal fees.

Subsidization is a particularly important issue for small businesses that generate hazardous waste.†† Small businesses often cannot pay the full cost of disposal and often have limited ability to treat and dispose of waste on site. The contribution of small businesses to some countries' hazardous waste problem is not trivial. In Hong Kong, for example, analysts have estimated that generators producing less than 100 kilograms of hazardous waste per month constituted 90% of hazardous waste generators, responsible for 50% of Hong Kong's hazardous waste.[4] Likewise, 98% of Thailand's factories are considered small.[5] Although the question of subsidies affects large businesses' *willingness* to pay for proper hazardous waste treatment and disposal, it raises the additional question of small businesses' *ability* to pay. It is not surprising, then, that countries with a significant small business problem, like Hong Kong, would choose to subsidize disposal fees.

**In Hesse, Bavaria, and Denmark, capital subsidies served to lower disposal fees in the early years of operation. Currently, though, disposal fees in all three areas are not considered to be subsidized and are regarded as quite high by European standards. After twenty-five to thirty years, the initial capital subsidies are largely irrelevant to the costs passed on to generators through fees today. However, the issue of subsidies may be more complicated than that. Generators in these three areas are required to obtain special permission to treat and dispose of hazardous waste outside their borders. This gives local TSD facilities a certain amount of monopoly power over those wastes that cannot be recycled or treated in other ways. In the absence of a competitive market for waste, these TSD facilities have more freedom to raise prices above actual costs—making the real effects of any capital subsidies on disposal fees much more uncertain.

††We use small businesses rather than small generators because large companies can—depending on their line of business—be small generators.

A Range of Approaches, A Range of Results

How effective have the various approaches to ownership and operation, incentives, and subsidies been in establishing a sound hazardous waste management system in each country? By effectiveness we mean, first, that treatment and disposal facilities are built and financially stable (so that capacity will remain available) and, second, that the majority of hazardous waste is properly managed. The discussion that follows looks at selected experiences in three groups, the private model (United States, Indonesia, and Malaysia), the public model (Denmark and Bavaria's ZVSMM facility), and the public–private model (Thailand, Alberta, and Hong Kong).

Private Model

The private model has worked well in the United States, in terms of both having adequate capacity and getting generators to use licensed hazardous waste TSD facilities. However, this has occurred against the backdrop of a strong enforcement regime, civil and criminal penalties for violations, and the threat of Superfund liability. The U.S. approach has also led to a largely disaggregated system of environmental treatment and disposal. Over 95% of hazardous wastes are disposed of on site in the United States. Because the bulk of these wastes are liquid wastes, they are either injected into "deep wells" or (after treatment) discharged to a publicly owned treatment plant.

The private-sector approach has, to date, not been as successful in Indonesia, and its prospects in Malaysia are open to question. With large land areas and relatively large industrial sectors, the HWM challenges in these countries are great. HWM policies are clearly aimed at establishing facilities to deal with large generators who could conceivably pay the full cost of disposal, rather than deal with the entirety of the hazardous waste problem. As a result, existing treatment and disposal capacity in both countries only begins to address the volume of waste generated. Even at these facilities, however, signs are not promising.

At the PT Prasadah Pemunahan Limbah Industry facility, Indonesia's single commercial HWM facility, waste volumes have been well below capacity, leading to low profitability. A relatively weak regulatory and enforcement system and limited public awareness of HWM issues, coupled with high disposal fees charged by the private owner, contribute to these problems. Indonesia's environmental regulatory agency, BAPEDAL, has made some progress in using its limited resources to focus on the approximately 1,000 large generators on Java and, through threats of enforcement and compliance assurance, in getting them to use the facility. The small size of BAPEDAL's hazardous waste staff and a political culture that favors voluntary compliance programs, however, have limited the agency's ability to enforce HWM regulations among generators. It is only with the passage of environmental legislation in 1997 (UU 23/1997) that BAPEDAL has had the authority to issue significant administrative penalties, making the threat of enforcement more credible.

In Malaysia, it is too early to tell how successful the first modern TSD facility—recently opened by the Kualiti Alam company—will be. At least one analysis has predicted that large companies faced with storage pressure will create a significant amount of business for the facility in spite of a weak regulatory system.[6] It is clear that the government's opposition to providing risk-reducing incentives to private investors contributed to delays in developing facilities, thus slowing the maturation of a functioning regulatory system. The opening of the new TSD facility lagged nearly ten years behind the promulgation of regulations, in part because the government and private firms were unable to negotiate various risk-sharing incentives to mutual satisfaction. In the late 1980s and early 1990s, two private firms withdrew from projects over these issues. The firm running the one commercial facility operates without any direct financial support or guarantees from the government, although it has been granted an exclusive license to build, operate, and maintain a centralized and integrated facility for fifteen years.[7]

Public Model

Denmark, with its Kommunekemi facility, and Bavaria, with its ZVSMM facility, took a mainly public-sector approach to financing. Consistent with the advantages of public-sector ownership discussed earlier, these facilities have provided considerable stability in terms of the availability of HWM capacity—Kommunekemi continues to operate after twenty-five years and ZVSMM, after thirty-two years. In Denmark, Kommunekemi still handles approximately 65% to 70% of the country's hazardous waste. Similarly, in Bavaria, according to analyses in the late 1980s, the ZVSMM facility and the state's other public–private facilities managed a much higher percentage of waste than what is treated on site. This extensive use of these facilities gave authorities a very complete picture of hazardous waste generation and management in the state.[8] At the same time that these public companies have provided predictable capacity, they appear to have avoided some of the bureaucratic entanglements that have dis-

credited public-sector approaches elsewhere. Both are considered to be efficiently operated and technically sophisticated operations.

Both Denmark and Bavaria subsidized disposal fees early on by financing
the construction of facilities with government funds. In both cases, local governments created these organizations by joining together to own and operate
hazardous waste treatment facilities. Federal (Denmark) or state (Bavaria) government provided some construction funds that covered a portion of capital
costs. The remainder of these costs, as well as operating costs, was recovered
through disposal fees. An analysis of the German system in the late 1980s con-
cluded that the ability to subsidize disposal fees in the early years—as was done
by Bavaria's ZVSMM—was an important component of program success.[9] Also
in Bavaria, the extensive collection system and subsidized treatment and disposal
allowed the state to require compliance by small and medium-sized generators,
rather than exempt them as in the United States.

Fees are not currently regarded as subsidized in either Denmark or
Germany, and in Denmark, at least, the Kommunekemi facility has been quite
profitable. One of the keys to Kommunekemi's success may have been the decision to operate as if it were a private company while still reaping the benefits of
public ownership. Indeed, in 1995, Kommunekemi reported an after-tax profit
of nearly US$4.5 million.[10] At the same time, its public status allowed it to
receive an initial interest-free loan from the national government with a ten-year
repayment deferral.

Whether the success of the public-sector approach taken by Denmark's
Kommunekemi and Bavaria's ZVSMM could be repeated in other countries is
open to question. As in the United States, the performance of facilities in these
countries has been against a backdrop of effective regulation and enforcement.
Denmark and, to a greater extent, Bavaria's Mittelfranken district, which ZVSMM
serves, are small and homogenous. The public companies benefited from long
traditions of public- and private-sector cooperation, local management, and relatively ordered and law-abiding societies.

Public–Private Model

Thailand, Canada's Province of Alberta, and Hong Kong all took approaches
that combined the public and private sectors in HWM facility ownership and
operation. Consistent with the idea that this approach gives countries maximum flexibility in deciding how to develop and charge for facilities, each of
these countries took a different approach to defining public- and private-sector

responsibilities, providing incentives to the private sector, and subsidizing (or not) disposal fees.

Thailand, although it has used public–private joint ventures to build and operate TSD facilities, has largely simulated a private market in hazardous waste management by charging full disposal fees (covering even construction costs). Government policy has been to award operation and management contracts to private-sector (or public–private) firms, which are then responsible for shouldering the operating risk and recovering the costs of collection, transportation, and landfilling from generators who use the facilities. At the country's Samae Dam facility, for example, the Ministry of Industry recovers capital costs by charging the operator a rental fee and a royalty fee based on the quantity of waste processed. The operator recovers these costs, as well as operating costs, through fees charged to those who use the facility.[11]

To date, there has not been enough hazardous waste treatment and disposal capacity available in Thailand to handle the estimated waste volumes in the country. In 1995, it was estimated that only 10% to 20% of hazardous waste generated in Thailand was being treated at an approved facility. Since that time, construction of an additional facility (at Rayong) has added 110,000 to 140,000 tons per year of capacity, but estimates of hazardous waste generated in the country range up to 1.5 million tons per year.[12] Even so, Thailand's two TSD facilities have been operating under capacity.[13] Many large producers—particularly those distant from the two commercial facilities—stockpile or treat wastes on site in uncontrolled facilities.[14] Subsidized transportation could have solved some of this problem. Many small-scale waste producers have insufficient access to collection, treatment, and disposal facilities and either stockpile waste or dispose of it in waterways or uncontrolled dumps.

Unlike any of the other areas we studied, Hong Kong and Alberta subsidized capital *and* operating costs in order to lower disposal fees. Originally, the Swan Hills facility in Alberta was owned by a public–private joint venture and operated by the private partner. Before privatization in 1996, Alberta subsidized approximately 40% of capital and operating costs at Swan Hills, savings that were passed on to customers of the facility through lower disposal fees. Progress in building an effective HWM system was not without its stumbling blocks, however. At Swan Hills, there was too much capacity in some process components and not enough in others—most notably, there was inadequate capacity to incinerate organic wastes from abandoned sites around the province.[15] Also, a high level of government subsidization stretched an already deficit-ridden provincial budget and played a large role in the ultimate decision to privatize the facility.

In Hong Kong, a public–private approach coupled with large subsidies has been relatively successful—at least compared with the other developing countries we examined—in encouraging proper waste treatment and disposal. In the first years of operation of its Chemical Waste Treatment Center (CWTC), Hong Kong fully subsidized disposal fees; in 1995, the government raised fees to 20% of operating costs. Hong Kong's approach was largely the result of the types of hazardous waste generators with which it was faced, since most hazardous waste came from small generators with little ability to treat their own waste or pay for proper disposal. The small size of the island and intense population density also made a centralized approach possible.

Waste volumes in Hong Kong have been consistent with capacity, and waste coming to the facility represents most of the waste thought to arise in the area. The transition to a less subsidized approach has been started, and anecdotal evidence suggests that the rise in disposal prices has spurred waste reduction among some large alkali waste producers.[16] Some analysts, however, believe that the rise in prices is now a barrier to small generators who are not willing to pay the cost.[17] For the most part, however, Hong Kong knows who the generators are; environmental officials report that most of the generators thought to exist have been identified and registered.

Lessons Learned

The eight countries we examined demonstrate a wide range of approaches to building and operating HWM facilities. With few examples of any one approach, it is difficult to draw general conclusions of what works and why. In a general sense, it is also hard to define success. We have defined success in terms of encouraging sound waste management practices and ensuring the financial stability of TSD facilities. Others, however, might define success in terms of minimizing public expenditures on hazardous waste management. In developing the lessons learned for this section, we assume that the *first* goal of any hazardous waste regulatory system is to control environmentally harmful waste disposal. Attention to controlling public expenditures—that is, shifting the financial burden off taxpayers and onto generators—becomes more important over time as regulatory programs mature and could be considered a goal for mature hazardous waste programs. Based on our examination of these countries, we reached three important conclusions.

1. There is no single "proper" approach to HWM facility financing that will work in every country.

Typically, experts in hazardous waste extol the virtues of *their* country's financing approach. Yet, our research shows that no one model—private, public, or a mix—is clearly superior to the others in all cases. The United States, Canada, Denmark, and Germany each have built effective HWM programs using a variety of approaches.

Rather than adopting a standard approach, each country should select a financing approach that is tailored to its circumstances (industrial profile, geography, government resources, and capacity), the effectiveness of its regulatory system, and its general policy objectives. For example, the public-sector approach taken by Hong Kong would not necessarily be appropriate for Indonesia. Hong Kong is small, transportation is a minor issue, small generators are a major part of the hazardous waste problem, and its per capita income is greater than that of Canada. Indonesia is vast and populous, transportation is a major issue, many facilities will ultimately be needed, and its per capita income is twenty-three times lower than that of Hong Kong. By the same token, what was right for the United States may not have been right for Indonesia either. The United States' strong legal system, adversarial regulatory climate, and cleanup liability laws—all of which are important elements of the country's hazardous waste regulatory system—are largely absent in Indonesia (and many other developing countries for that matter).

2. In countries where there is not yet a culture of compliance, the financing approach matters.

Although there is no standard approach to financing hazardous waste facilities, we can generalize about the need for subsidies in countries where enforcement and compliance are weak—a situation typical of the early years of regulatory programs. In this environment, treatment and disposal facilities usually "compete with the river" where the cost of disposal is zero. As countries make the transition from an unregulated environment, some form of carrot or stick (or both) needs to be used to change the behavior of hazardous waste generators. When the regulatory stick is weak—as in all of the developing countries we examined—financing models that allow countries to offer the carrot of subsidized disposal fees are likely to lead to more effective systems in the short to medium term.

Of the programs we studied, those that subsidized hazardous waste disposal in the early years—such as Hesse (Germany), Bavaria (Germany), Denmark, Hong Kong, and Alberta (Canada)—all had some degree of public ownership in HWM facilities. Although not represented in the countries we looked at, there may be ways to subsidize disposal fees even if facilities are privately owned and operated, through a voucher system, for example. The important issue is not who owns and operates facilities, but what generators must pay to use them.

3. Disposal fee subsidies are a viable transitional strategy for encouraging proper waste disposal.
Subsidized disposal fees can be viewed as a *transitional* strategy for pursuing the long-term goal of building a culture of compliance. That is, it is common sense to phase in more stringent and costly HWM requirements over time. Getting generators in the habit of using an HWM facility is an important first step in such a process. Such an approach also begins to provide data on the amount and type of hazardous waste generated, information that can help accelerate and strengthen the development of an effective regulatory and enforcement program.

A number of countries have shown that transitions from a subsidized approach toward a more market-driven approach are possible. Denmark and Germany's Bavaria and Hesse, all of which had subsidized disposal fees, now charge very high prices for disposal. The Swan Hills facility in Alberta, Canada, operated with subsidized disposal fees for nine years while the province's regulatory system was maturing, and then the facility was privatized in 1996. Similarly, Hong Kong's CWTC operated for two years with fully subsidized disposal fees, until they were raised to 20% of disposal costs in 1995.

Endnotes

1. The discussion of advantages and disadvantages of each approach comes from information provided in Environmental Resources Management (ERM), *Public/Private Sector Cooperation in the Provision of Hazardous Waste Management Facilities* (London: ERM, 1994), p. 9; and in Dames & Moore, *Draft Report: Feasibility and Basic Design Studies, Hazardous Waste Treatment and Disposal Facility in Jabotabek Area, Industrial Efficiency and*

Pollution Abatement Project, Republic of Indonesia, submitted to the World Bank (Bethesda, Maryland: Dames & Moore, 1993), Sections 6.3-1 to 6.3-4.

2. ERM, *Public/Private Sector Cooperation,* p. 30.

3. Data for each country in Table 3-1 are taken from the following sources:

Hong Kong: Patrick Heininger, "Solving the Hazardous Waste Problem in Developing Countries," unpublished manuscript (Singapore: Waste Management International, January 1998); E-mail from R.C. Rootham to T. Beierle, May 11, 1998, and to K. Probst, December 5, 1998; ERM, *Public/Private Sector Cooperation,* pp. C8–C9.

Thailand: Communication between Satit Sanongphan (United States–Asia Environmental Partnership, Thailand) and T. Beierle, June 2, 1998; E-mail from Sombat Sae-Hae to T. Beierle, May 27, 1998, and June 12, 1998.

Indonesia: Heininger, "Solving the Hazardous Waste Problem"; ERM, *Public/Private Sector Cooperation,* pp. A11–A12.

Malaysia: E-mail from Michael Hansen (Kualiti Alam, Malaysia) to T. Beierle, May 21, 1998; E-mail from Ibrahim Shafii (Department of the Environment, Malaysia) to T. Beierle, May 14, 1998; ERM, *Public/Private Sector Cooperation,* pp. B5-B8; E-mail from Lim Thian Leong to K. Probst, December 9, 1998.

4. Ibid., p. 8.

5. E-mail from Sombat Sae-Hae (Thailand Development Research Institute) to K. Probst, December 4, 1998.

6. ERM, *Public/Private Sector Cooperation,* p. B9.

7. E-mail from Lim Thian Leong (Center for Environmental Technologies, Malaysia) to T. Beierle, May 8, 1998, and to K. Probst, December 9, 1998.

8. Joanne Linnerooth and Allen V. Kneese, "Hazardous Waste Management: A West German Approach," *Resources* (Summer 1989), pp. 8–9.

9. Joanne Linnerooth and Gary Davis, *Hazardous Waste Policy Management— Institutional Dimensions* (Laxenburg, Austria: International Institute for Applied Systems Analysis, 1984), p. 30.

10. "Net Profits up 43% at Kommunekemi in 1995," *Haznews* (July 1, 1996).

11. Boonyang Lohwongwatana, Teerapon Soponkanaport, and Aioporn Sophonsridsuk, "Industrial Hazardous Waste Treatment Facilities in Thailand," *Waste Management and Research,* Vol. 8 (1990), pp. 129–134.

12. E-mail from Sombat Sae-Hae to T. Beierle, May 27, 1998; and Thailand Development Research Institute (TDRI), *The Monitoring and Control of Industrial Hazardous Waste: Hazardous Waste Management in Thailand* (Bangkok: TDRI, 1995), p. 18.

13. Fax from Suriya Supatanasinkasen (General Environmental Conservation Company, Thailand) to T. Beierle, January 14, 1999.

14. Communication between Satit Sanongphan and T. Beierle, June 2, 1998.

15. Barry G. Rabe, *Beyond NIMBY: Hazardous Waste Siting in Canada and the United States* (Washington, D.C.: Brookings Institution, 1994), p. 88.

16. E-mail from R.C. Rootham (Local Control Office [Territory East], Environmental Protection Department, Hong Kong) to T. Beierle, May 11, 1998.

17. E-mail from Keith Gilges (United States–Asia Environmental Partnership, Hong Kong) to T. Beierle, June 9, 1998.

4. Other Important Issues and Areas for Further Research

In the course of our research, several issues arose that are not directly related to program evolution or facility financing but are, nonetheless, important elements in developing a hazardous waste management program. We briefly describe them here and then conclude our report with recommendations for areas warranting additional research.

Other Important Issues

1. One very important resource in a hazardous waste management program is *information*—about who is generating waste, what quantities and types are being generated, and where it is going. One of the most important ancillary benefits of Hong Kong's interim arrangements and Malaysia's early system of registration prior to the opening of Kualiti Alam's facility was information on waste generation.[1] In Hong Kong, the few years of free disposal at the Chemical Waste Treatment Center (CWTC) allowed the government to collect information on generators. When business dropped off at CWTC after prices were raised, the regulators knew which facilities to visit to find out why.

In general, the publicly funded programs we examined tended to have better information on waste generation. They attracted a greater number of small and medium-sized generators by subsidizing disposal fees. Also, many of these publicly funded facilities are centralized, making data collection easier. In Bavaria, for example, there were significant information advantages arising from the near total use of one facility by industry. As of the late 1980s, at least, Bavarian authorities had a relatively complete picture of hazardous waste generation and management in the state.[2] By contrast, data on hazardous waste generation in the United States are quite poor.[3] Of course, the sheer size of the United States makes information collection more difficult. In addition, national U.S. data are hard to come by because much of the hazardous waste information is collected at the state level. However, the problems are aggravated by the disaggregated nature of the system, including the extent of on-site treatment.

2. Regardless of whether facilities are financed publicly or privately, accurately estimating needed future capacity is difficult.
Clearly, accounting for geography, the type of waste, and the nature of generators matters. Deciding which industries to focus on first matters. But, predicting demand for waste treatment and disposal was tricky in all the countries we examined. Indonesia is a case in point: analysts had identified potential customers for a facility, but those customers decided not to use it. Demand is hard to predict before a regulatory regime is in place, but it is also very hard to predict once generators must begin paying for hazardous waste disposal. Generators quickly find creative ways (some environmentally sound and some not) to reduce the cost of hazardous waste treatment and disposal. In addition, over time, demand changes as countries' industrial makeup changes, and technologies change as economies grow and evolve. For example, Germany saw a capacity crunch in the late 1980s turn into a capacity glut in the early 1990s, at least in part because high disposal fees led generators to reduce waste generation or seek alternate means of disposal.[4]

3. Planning for hazardous waste infrastructure must account for the geography of hazardous waste generation and the cost of transportation from generators to treatment and disposal facilities.
Often, the industries that generate hazardous waste are concentrated geographically. Even if a country has treatment and disposal facilities, if distances between generators and facilities are too great and transportation costs too high, the result can be similar to a situation in which no treatment and disposal

capacity exists. Thailand is a good example. Although the country has two major facilities for the treatment and disposal of hazardous waste, those facilities serve mainly Bangkok and the central and eastern regions. For many generators, transportation costs are prohibitive.[5] Instead, there has been considerable on-site storage, even to the extent that companies are reportedly buying warehouses or property simply to store waste.[6]

4. There are a number of nonmarket approaches that can be (and have been) used to encourage waste reduction and recycling.
One of the principal arguments against subsidizing disposal fees is getting the right financial incentives to encourage waste reduction and recycling. If disposal fees are subsidized, theory would argue, firms would be encouraged to overproduce waste relative to a theoretical optimum. However, countries have used a number of nonprice mechanisms to encourage waste reduction and recycling, both in addition to and in lieu of charging the full cost of disposal. These strategies include:

- *Laws and policies to encourage waste minimization.* The United States' Pollution Prevention Act of 1990, Germany's 1996 Recycling and Waste Management Act, Denmark's 1984 Recycling and Waste Reduction Act, and Canada's 1988 revision of the national Environmental Protection Act are all examples of laws that emphasize waste reduction, reuse, and recycling as alternatives to disposal.
- *Public information.* The United States' Toxic Release Inventory encourages firms to reduce pollution through the public disclosure of information on their environmental performance. Although Indonesia's PROPER system does not currently include information on hazardous waste management, it is another example of companies' environmental performance being widely publicized.
- *Superfund-type liability.* In the United States, the potential for cleanup liability is a powerful motivator for waste reduction and proper waste management.
- *Subsidies or requirements for waste reduction equipment.* In Denmark, concerns that subsidizing disposal fees would discourage waste minimization led the Danish government to implement a program to subsidize up to 15% of capital costs for in-plant technology to reduce waste.[7] In

Germany's state of Hesse, the government counteracted the effect of subsidized prices on discouraging waste minimization by instituting a policy
requiring all new industrial plants to include best-available-control technology for reducing pollution and recycling wastes.[8]

5. Siting hazardous waste facilities is always controversial.
In the United States, public opposition to hazardous waste facilities has led to
an almost universal halt in off-site facility construction since the mid-1980s. In
Thailand, a major constraint on building additional facilities has also been public opposition. The construction of the hazardous waste treatment and disposal facility in Rayong was delayed in the mid-1990s due to continued protests
from the local population, which ultimately led to the re-siting of the facility.
The Kualiti Alam project in Malaysia also saw significant delays, partly due to
local opposition.[9] In Ontario, Canada, a fifteen-year effort to site a facility
recently failed.[10] Research in Canada and the United States suggests that active
public involvement in the siting process can reduce delays from opposition.[11]
Thus, whether a country is considered "developed" or "developing," public
opposition to siting hazardous waste management facilities can be a major
obstacle to bringing additional hazardous waste treatment and disposal capacity on-line.

Areas for Further Research

In the course of our work, it became clear that a number of important issues
warrant further study. We begin with the most important issue and move to
other interesting areas of research.

1. What can be done to create a culture of compliance?
It is no surprise that we conclude that a culture of compliance is necessary to
implement a successful hazardous waste management program. It is difficult to
imagine any comprehensive hazardous waste regulatory system functioning in
a situation where corruption is the norm and where rules and regulations are
routinely ignored and violated. To be effective, regulatory programs need a government with public credibility. Further research needs to combine the wealth
of academic literature on this issue with the practical experiences of government officials and regulated entities in a range of countries.

2. What have been the experiences in other parts of the world, such as Eastern Europe, Latin America, and Africa? Are the findings for countries in these regions consistent with those reached in our research?

As we noted in Chapter 1, the developing countries we studied are all in Southeast Asia. It would be extremely useful to examine a more geographically diverse set of developing countries, and perhaps a larger group of developed countries as well, to test our lessons learned against a wider range of experiences. Different areas of the world face different internal and external financial and social pressures. Thus, it would be helpful (and interesting) to look at the experiences of Eastern Europe, Latin America, and Africa.

3. How should subsidies be structured to "get the incentives right" for hazardous waste generators, investors, and the government?

Our work documents that subsidies are often part of the mix of tools used to get hazardous waste management facilities on-line and fully utilized. Sometimes they are effective, and sometimes they are not. More work on this issue is needed in a broader range of countries, and in more detail, to be able to reach solid conclusions on how subsidies (and other pricing strategies) would best be structured for different objectives such as encouraging investment in facilities, encouraging generators to use facilities, and encouraging waste reduction.

4. How should public–private ventures be structured? What arrangements work best in different situations?

In most of the countries we studied, the public- and private-sector approaches were combined in interesting ways. Other researchers have noted that public–private joint ventures are the most common approach for building and operating hazardous waste management facilities worldwide. More research, both theoretical and applied, is needed to understand the advantages of public- and private-sector partnerships in developing hazardous waste management infrastructure.

Endnotes

1. Environmental Resources Management (ERM), *Public/Private Sector Cooperation in the Provision of Hazardous Waste Management Facilities* (London: ERM, 1994), pp. 27–28, 35.

2. Joanne Linnerooth and Allen V. Kneese, "Hazardous Waste Management: A West German Approach," *Resources* (Summer 1989), p. 9.

3. J. Clarence Davies and Jan Mazurek, *Pollution Control in the United States: Evaluating the System* (Washington, D.C.: Resources for the Future, 1998), pp. 79–80.

4. Fax from Gerhard Smetana (Umweltbundesamt, Germany) to T. Beierle, May 12, 1998.

5. Communication between Satit Sanongphan (United States–Asia Environmental Partnership, Thailand) and T. Beierle, June 2, 1998.

6. Thailand Development Research Institute (TDRI), *The Monitoring and Control of Industrial Hazardous Waste: Hazardous Waste Management in Thailand* (Bangkok: TDRI, 1995), p. 15; and communication between Satit Sanongphan and T. Beierle, June 2, 1998.

7. Bruce Piaseki and Gary A. Davis, "A Grand Tour of Europe's Hazardous-Waste Facilities," *Technology Review* (July 1984), pp. 27–28.

8. Ibid., p. 32.

9. "Hazardous Waste Management in Industrializing Countries," *Haznews*, No. 59 (February 1993).

10. Environment Canada, *Status Report on Hazardous Waste Management Facilities in Canada—1996* (Ottawa, Ontario: National Office of Pollution Prevention, February 1998), p. 9.

11. This is one of the principal conclusions of Barry G. Rabe, *Beyond NIMBY: Hazardous Waste Siting in Canada and the United States* (Washington, D.C.: Brookings Institution, 1994).

Appendix A: Country Profiles

The eight countries we examined are a highly diverse group. Four are Western developed nations that industrialized around the turn of the century, and four are Asian developing countries that industrialized largely over the last two decades. Three of the countries—the United States, Canada, and Indonesia—have land masses greater than one million square kilometers, while Hong Kong* covers just under 1,100 square kilometers. There is a great diversity of wealth as well. Germany, for example, has a gross national product of US$2.3 trillion, whereas Malaysia's economy is one-thirtieth that size. Thailand's economy is roughly the same as Denmark's, but it has over ten times the population and land area.[1]

In spite of these many differences, hazardous waste management (HWM) programs in these eight countries share some common elements, many of which are highlighted in Chapters 2 and 3. In this appendix, we briefly summarize key aspects of each country's HWM program, including its evolution, the level of government with principal HWM responsibility, the number of commercial HWM facilities, and public and private roles in facility financing. These aspects are summarized in Table A-1.

---

*This report examines Hong Kong's approach before its reversion to China.

Table A-1. Summary of Country Characteristics

	Date of Comprehensive HWM Laws or Regulations	Level of Government with Main HWM Regulatory Authority	Source of Facility Financing
Developed Countries			
Germany	1972 (law)	State	Public/private
Denmark	1973 (laws)	National	Public
United States	1976 (law)	National	Private
Canada	1980 (law)	Provincial	Public/private
Developing Countries			
Malaysia	1989 (regs.)	National	Private
Hong Kong	1991 (laws)	National	Public/private
Thailand	1992 (laws)	National	Public/private
Indonesia	1994 and 1995 (regs.)	National	Private

Note: HWM, hazardous waste management.

We present the countries in the order in which they passed their main hazardous waste laws or regulations. Germany was the first of the eight countries to enact a law specifically addressing hazardous waste management, in 1972. Denmark followed in 1973, the United States in 1976, and Canada in 1980. In the developing countries we studied, laws and regulations specifically governing hazardous waste were passed in the late 1980s and early 1990s. Malaysia was the first of this group in 1989, followed by Hong Kong in 1991, Thailand in 1992, and Indonesia in 1994. (See also Chapter 2, Figure 2-1.)

Germany

Germany is a federal republic in which sixteen federal states (eleven western states and five "new" states of eastern Germany) have considerable control over environmental policy, including policies to manage hazardous waste. In the early years of hazardous waste program development, individual states were largely responsible for designing and implementing regulatory programs and determining how HWM capacity would be developed. In some of the states, the construction of centralized hazardous waste disposal facilities predated Germany's national hazardous waste law passed in 1972. In examining facility financing, we focus on two states, Bavaria and Hesse, where the public sector played a significant role in the construction and operation of HWM facilities.

Hazardous waste is referred to as "special waste" in Germany. Special wastes are those "generated by industrial, commercial, or public sources which by reason of their nature, condition, or quantity, constitute a particular hazard to health or the quality of air or water, or are particularly explosive or flammable, or contain or may lead to the development of pathogenic organisms."[2] Germany's hazardous waste regulations currently name 332 types of waste belonging to the category of special waste. Nuclear waste, wastewater, military wastes, and mining wastes are regulated separately from special waste. Waste oil has additional regulations that supplement those covering special waste.

According to 1993 data, Germany generates approximately nine million tons of hazardous waste annually.[3] As of 1992, sixteen landfills and thirty-one large incinerators comprised Germany's large-scale, commercial, off-site hazardous waste disposal capacity.[4]

The following sections discuss hazardous waste regulatory development and HWM facilities in Germany in greater detail. Table A-2 presents a time line of important dates in the development of the country's HWM program.

Program Development

Germany's 1972 Waste Disposal Act (WDA) provided the early legislative framework for all of the country's waste management. Although WDA did not refer specifically to "special waste," it distinguished from other solid waste those wastes that "may endanger health, air, or water, be explosive or inflammable, or contain or cause infectious diseases."[5] WDA delineated lines of regulatory authority; required the licensing and information disclosure of treatment, storage, and disposal (TSD) facilities; and laid out a cradle-to-grave control system with manifests and record keeping. Although the law established a federal framework for hazardous waste management, it gave the states discretion on implementation and enforcement. Two years later (1974), the federal Imission Control Law required licensing of hazardous waste generators. In the same year, the national environmental regulatory agency, Umweltbundesamt, was established.

The concept of hazardous waste introduced in the 1972 WDA was further defined by an administrative order in 1977 that specified a list of eighty-six specific wastes to be regulated (the list could be expanded by individual states). This number was increased to 332 in 1990. In 1978, the Waste Manifest Administrative Order established a manifest system for tracking hazardous waste from

Table A-2. Hazardous Waste Program Development in Germany

Year	Laws/Policy	Agency	Regulations	Facilities
1966				ZVSMM established in Bavaria
1968				*Through 1974:* ZVSMM opened landfill, treatment facility, and incinerator
1970				GSB established in Bavaria
1972	Waste Disposal Act			
1974	Imission Control Law	National environmental agency (Umweltbundesamt) established	*Through 1978:* Series of administrative orders on notification of waste generation, waste types, and manifests	HIM established in Hesse
1986	Waste Avoidance and Management Act			
1990			Technical Instructions on Hazardous Wastes	
1996	Recycling and Waste Management Act		Establishment of seven ordinances and technical instructions that currently govern hazardous waste	

Notes: GSB, Gesellschaf zur Entsorgung von Sondermull (Bavaria); HIM, Hessische Industriemull GmbH (Hesse); and ZVSMM, Mittelfranken Cooperative for Special Waste Management (Bavaria).

generation to disposal (amending the 1974 Administrative Order on the Notification of Waste). By 1985, the manifest system covered approximately 90% of
hazardous waste produced in Germany.[6]

In 1986, the Waste Avoidance and Management Act (WAMA) substantially revised the 1972 WDA. WAMA harmonized many of the state standards and
established a more centralized command-and-control hazardous waste regulatory system. Regulations governing the storage, treatment, incineration, and disposal of waste were passed in 1990 as the Technical Instructions on Hazardous
Wastes. A 1993 performance review by the Organisation for Economic Co-operation and Development (OECD) considered western Germany's industrial waste
standards, including those for incineration and landfilling, to be among the most
stringent in the world.[7]

In 1996, the Recycling and Waste Management Act (RWMA) superseded
WAMA. The new law focused on material recycling and extended product
responsibility. RWMA placed primary legal and financial responsibility for the
entire life cycle of products, including waste management, on producers themselves. The current HWM system is governed by seven ordinances and a technical instruction, all of which were put in place in 1996.

Hazardous Waste Management Facilities

Since the inception of the country's HWM program in the 1970s, Germany's
states have been differentiated by the extent to which the public sector has been
involved in the construction and operation of hazardous waste TSD facilities.
North Rhine-Westfalia, for example, left infrastructure development to the private sector, while Bavaria and Hesse—the two states on which this section
focuses—established TSD facilities with substantial public investment.

In Bavaria, two companies were established with public investment in the
late 1960s and early 1970s with the purpose of building and operating hazardous waste facilities. The Mittelfranken Cooperative for Special Waste Management (Zweckverband Sondermullplatze Mittelfranken, known as ZVSMM) was
a town- and county-owned cooperative established in 1966 to manage wastes
from Mittelfranken, an industrialized area of Bavaria. By 1974, ZVSMM had built
a hazardous waste landfill and a treatment and incineration facility, both wholly
owned by the towns and counties of Mittelfranken. Some capital costs for these
facilities were paid by the state government (either directly or through low-interest loans) and not recovered. At least into the early 1980s, these capital subsidies were not being recovered through disposal fees. Rather, disposal fees were

recovering only a portion of the total capital cost of the facilities, as well as operating costs and interest payments.[8]

Bavaria's second HWM company, the Association for the Management of Special Wastes (Gesellschaft zur Entsorgung von Sondermull, or GSB) was established in 1970. GSB is a public–private partnership, with the state owning 70% and a group of industries that generate hazardous waste owning the remainder. GSB serves all of Bavaria, except the Mittelfranken area, with its landfill, treatment, incineration, and recycling facilities. Like ZVSMM, in the early 1980s disposal fees charged by GSB were subsidized; they were not high enough to pay off the public-sector investment in the capital costs of the facilities.

Hesse is also served by facilities built and operated by a public–private partnership. The partnership was formed when a consortium of industries that had invested in the construction of hazardous waste treatment and disposal capacity ran into financial difficulty in the early 1970s and was bailed out by the state. The state first provided low-interest loans, then direct subsidies, for additional HWM capacity through the public–private partnership, which was formed in 1974. Twenty-eight percent of the partnership (Hessische Industriemull GmbH, or HIM) is owned by the public sector and 72% by the private sector. As in the Bavarian case, the publicly financed capital expenditures were used to subsidize disposal fees.[9]

There is considerable precedent in Germany for joint private-sector–public-sector funding of environmental infrastructure. Germany's approach to public policy decisionmaking entails a high degree of coordination among the public and private sectors as well. This model stresses consultation, negotiation, and cooperation among the large public sector, business trade associations, and trade unions. This cooperative approach provides the context for public financial incentives for environmental protection. Such incentives have been common practice.

Denmark

Denmark is the smallest of the developed countries we studied, with a population of 5.2 million (less than three-quarters that of New York City) and a land area of 43,000 square kilometers (approximately one-eighth the size of Germany). Its HWM system is considered to be extremely effective. The system's two main features are a highly decentralized process of inspection and oversight of facilities coupled with a highly centralized system for collecting, transferring, treating, and disposing of waste. Denmark's principal environmental laws delegate most implementation responsibilities to the country's 275 municipalities and, to some extent, its fourteen counties. In the mid-1970s, these local governments joined together to form the Kommunekemi, a centralized HWM facility that still processes much of Denmark's hazardous waste through incineration or treatment and landfilling.

The terminology for hazardous waste in Denmark distinguishes between "waste oils" and "waste chemicals." The list of wastes that fall under these designations is consistent with European Community directives.[†] Both types of wastes are referred to here as hazardous waste.

In 1995, Denmark generated approximately 250,000 tons of hazardous waste.[10] Much of this waste ended up at Kommunekemi. Between 1993 and 1995, the facility received annually between 90,000 and 105,000 tons of hazardous waste and between 15,000 and 20,000 tons of contaminated soil.[11]

The following sections discuss hazardous waste regulatory development and HWM facilities in Denmark in greater detail. Table A-3 presents a time line of important dates in the development of the country's HWM program.

Program Development

Denmark's main environmental law for controlling industrial pollution was the 1973 Environmental Protection Act. It was a framework law that left to the Ministry for Environment and Energy (prior to 1994, the Minsistry of the Environment) the task of developing specific rules and regulations through statutory orders. The 1973 law established a system, which is still in use, of listing individual firms, designated as "particularly polluting enterprises," that fall within certain industrial sectors. Listed companies are then subject to regulatory supervision by municipalities or counties and must comply with a standard set of requirements for their sector. All new enterprises in these designated sectors are added to the list and are required to apply for environmental approval. Currently, the list covers sixty industrial sectors and includes 10,000 individual enterprises. The Environmental Protection Act was amended numerous times over the years until a complete revision was enacted in 1991.

Denmark's primary chemical waste law, the Act on the Disposal of Oil and Chemical Waste, was passed in 1973 along with the Environmental Protection Act. In 1983, Denmark passed its first law regarding the cleanup of chemical

---

[†]These directives are 91/689 for waste chemicals and 75/439 and 87/101 for waste oils.

Table A-3. Hazardous Waste Program Development in Denmark

Year	Laws/Policy	Agency	Regulations	Facilities
1971		Ministry for Pollution Control established		
1972		Environmental Protection Agency established		
1973	Environmental Protection Act Act on the Disposal of Oil and Chemical Waste	Ministry of the Environment established		
1975				Kommunekemi opened
1983	First law concerning cleanup of hazardous waste disposal sites			
1984	Recycling and Waste Reduction Act			
1987			*Through 1990:* Municipalities completed lists of regulated facilities	
1990	Contaminated Sites Act			
1991	Revision to Environmental Protection Act			
1994		Merger of ministries of environment and energy created Ministry of Environment and Energy		

waste disposal sites and in 1990 expanded the program through the passage of the Contaminated Sites Act. In 1984, Denmark passed the Recycling and Waste Reduction Act, which focused on waste reduction and established Denmark's waste control hierarchy as clean technology at the top, followed by recycling, incineration, and disposal.

Environmental authority in Denmark is shared among the municipal, county, and national levels. Municipalities—and, increasingly, counties—have the majority of responsibility for inspecting and sanctioning facilities, while the national authorities set standards and settle disputes between industries and local regulators. The system of supervision and inspection of listed enterprises for compliance with hazardous waste requirements began to really function around 1987. Between 1987 and 1990, municipalities completed registers of enterprises over which they have supervisory authority. [12]

Elected councils for Denmark's 275 municipalities have been the main administrative bodies for environmental matters—including hazardous waste— since the inception of Denmark's environmental laws in 1973. They have primary responsibility for inspections and all environmental regulations not specifically the responsibility of higher public bodies. On average, municipalities inspect facilities once every three years.

Denmark's fourteen counties, due to their larger staff size and greater technical capacity, have become increasingly responsible for implementing some aspects of environmental law. In particular, they have had increasing responsibility for inspecting and sanctioning more complex facilities. Recently, counties have taken the lead in administering policies regarding contaminated sites.

At the national level, the Ministry of Environment and Energy oversees the national Environmental Protection Agency and develops the rules (statutory orders) that implement environmental laws. The Environmental Protection Agency develops influential environmental management guidance for counties and municipalities and handles appeals of environmental rulings made at the municipal and county level. Although the Ministry of Environment and Energy has full power to direct the activities of the Environmental Protection Agency, it does not have such power over municipalities and counties, which retain their independence in carrying out regulatory activities.

Hazardous Waste Management Facilities

After passage of the Act on the Disposal of Oil and Chemical Waste in 1973, Denmark's municipalities and counties formed a consortium to finance the construction and operation of the Kommunekemi facility, the focal point of Denmark's HWM system. The construction of Kommunekemi was financed by an interest-free loan from the national government to the municipal consortium. It included a ten-year repayment deferral. The facility opened in 1975.

Most waste at Kommunekemi is incinerated, with the residue going to its associated landfill. Inorganic waste is treated and either landfilled or transferred to the city sewer system if heavy metal content does not exceed limits. Kommunekemi accepts the majority of the nation's hazardous waste through a network of twenty municipally owned and operated industrial waste collection and transfer stations, which, in turn, are linked to 275 household chemical drop-off stations. Kommunekemi is the only facility in Denmark authorized to receive hazardous waste for treatment and disposal; however, municipalities may give firms special exemptions to use or recycle some types of hazardous waste. Most of the waste receiving exemptions is waste oil to be used in district heating plants. The remaining wastes are mainly chemicals that can be recycled at facilities other than Kommunekemi. Private recycling firms have been established to handle this latter type of waste.

Kommunekemi was constructed with public funds and continues to operate as a publicly owned company. In expressed adherence to the "polluter-pays principle," however, Kommunekemi seeks to recover capital and operating charges through disposal fees, currently ranging from around US$400 per ton for ordinary chemicals to around US$4,000 per ton for difficult-to-treat wastes. In recent years, such disposal fees have resulted in annual revenues at the facility of approximately US$33 million.

The public-sector approach taken at Kommunekemi is consistent with Denmark's general approach to environmental investment. This approach typically has been characterized by significant public expenditure coupled with a variety of user fees, taxes, and levies. For example, US$18 million was committed to waste and recycling projects for the period 1993 to 1997. Municipalities typically undertake their solid (as opposed to hazardous) waste disposal facilities as joint ventures with private firms. As of 1992, all landfills must be publicly owned due to postclosure contamination concerns. At the same time that it finances infrastructure with public funds, Denmark has an extensive system of user fees to recover costs and influence behavior. These include service charges (for example, sewage disposal, water supply, refuse collection, and environmental authorization and supervision) and environmental levies (for example, energy, automobiles, and carbon dioxide).

United States

In the United States, hazardous waste management is primarily a private-sector responsibility. Firms generating, transporting, storing, treating, and disposing of hazardous waste are subject to strict regulations and strong enforcement. Laws making firms liable for the cleanup of contaminated sites provide a strong incentive for proper waste management.

Under U.S. Environmental Protection Agency (EPA) regulations, hazardous waste is defined as waste that exhibits certain characteristics (ignitability, corrosivity, or reactivity) or has the potential to leach a certain amount of toxic chemicals. In addition, EPA can designate specific wastes (referred to as "listed wastes") as hazardous and, therefore, subject to regulation. There are currently more than 500 listed wastes. Several states have their own definitions of hazardous wastes that are more inclusive than the federal definition. In addition, some state HWM programs regulate small-quantity wastes that are exempt under the federal program.

The U.S. Congress has specifically exempted certain wastes from regulation under the United States' major HWM law, the Resource Conservation and Recovery Act. These exemptions include wastes from mining, petroleum production, electricity generation, and small sources. Radioactive waste is regulated under the Atomic Energy Act rather than by the laws that cover hazardous waste (unless the radioactive waste is mixed with hazardous waste, in which case it is regulated under both radioactive and hazardous waste laws).

According to EPA statistics, nearly 20,000 generators in the United States produced approximately 279 million tons of hazardous waste in 1995, the latest date for which comprehensive statistics are available. These data include 267 million tons of wastewater (or 96% of the total), which is considered hazardous waste under U.S. definitions.

The following sections discuss hazardous waste regulatory development and commercial HWM facilities in the United States in greater detail. Table A-4 presents a time line of important dates in the development of the country's HWM program.

Program Development

Before passage of the Solid Waste Disposal Act (SWDA) in 1965, waste management was considered primarily a local concern. After the passage of SWDA, more than a decade passed before Congress enacted significant legislation

Table A-4. Hazardous Waste Program Development in the United States

Year	Laws/Policy	Agency	Regulations	Facilities
1965	Solid Waste Disposal Act			
1970	Resource Recovery Act			
1973		U.S. EPA established		
1976	Resource Conservation and Recovery Act (RCRA)	U.S. EPA Office of Solid Waste established		
1978			Through 1983: RCRA "base program" regulations promulgated	
1980	RCRA Amendments; Comprehensive Environmental Response, Compensation, and Liability Act (CERCLA)		new requirements (1984 amendments severely limit land disposal of hazardous wastes)	Through 1985: Many TSD facilities shut down rather than comply with
1984	RCRA Amendments (Hazardous and Solid Waste Amendments)			
1985			Through 1990: Regulations implementing 1984 RCRA Amendments (land disposal restrictions and corrective action requirements)	
1986	CERCLA Amendments			
late 1980s				There has been a virtual halt to construction of commercial TSD facilities mainly as a result of opposition to siting new facilities
1990	Pollution Prevention Act			

Notes: U.S. EPA, U.S. Environmental Protection Agency; TSD, treatment, storage, and disposal.

addressing hazardous waste. Two major laws—the Resource Conservation and Recovery Act (RCRA) and the Comprehensive Environmental Response, Compensation, and Liability Act (CERCLA), commonly known as Superfund—form the foundation of the U.S. hazardous waste management system. CERCLA, which was enacted in 1980 and significantly amended in 1986, deals with the cleanup of sites contaminated with hazardous substances.

RCRA was passed in 1976 as a body of amendments to SWDA and was amended in 1980 and 1984. To implement the law, Congress created (in 1976) an Office of Solid Waste within EPA. (EPA is the country's primary regulatory body, established in 1973.) RCRA Subtitle C deals specifically with the definition, handling, storage, treatment, and disposal of hazardous waste (which, rather confusingly, is technically referred to as "solid waste.")

From 1978 to 1983, EPA established what are known as RCRA's "base program" regulations. Many of these were promulgated in 1980, including rules concerning the record keeping, reporting, labeling, and storage requirements with which generators must comply and a manifest system for the transfer of waste. Final rules on performance standards for TSD facilities were promulgated in 1982.

At the time of RCRA's first major set of amendments in 1984—eight years after its initial passage—the program still was not fully implemented. These amendments required state plans for regulating and remediating underground storage tanks, placed significant restrictions on the land disposal of hazardous waste, and required environmental cleanup at facilities managing hazardous waste (referred to as the "corrective action" program). Partly due to frustrations with EPA's slow pace in implementing RCRA, Congress made the 1984 amendments the "most detailed and restrictive environmental requirements ever legislated."[13] Most of the regulations implementing RCRA's 1984 amendments were promulgated between 1985 and 1990, although some are still being written in the late 1990s. Most analysts agree that the hazardous waste regulatory system was, for the most part, fully operational by the late 1980s.

Under RCRA, EPA can authorize states to implement their own HWM programs in lieu of the federal program, if the state program meets or exceeds federal requirements. Forty-seven states, Guam, and the District of Columbia have EPA-authorized programs, all with some degree of federal funding.

According to a 1996 OECD review, the combination of direct sanctions (RCRA) and the threat of cleanup liability (CERCLA) has led to a regulatory regime in the United States in which "potential violators can be reasonably sure that they will be sought out and made to redress any proven violations."[14] However, the OECD review also found the system to be overly complex and expensive. The complexity of RCRA has also resulted in noncompliance, particularly among smaller businesses that do not have the capacity to fully understand and comply with the rules.

Recent legislation and policies have added waste minimization as a major component of the nation's HWM program. The 1990 Pollution Prevention Act and voluntary policies, such as EPA's 33/50 program, are designed to provide incentives for minimizing solid and hazardous waste production. Public pressure, public right-to-know programs, and fear of Superfund liability have also provided incentives to reduce the use of hazardous substances.

Hazardous Waste Management Facilities

In the United States, the construction and operation of HWM facilities are solely the responsibility of the private sector, operating under a strict regulatory regime. Contaminated site cleanups are also paid for mainly by the private sector.[15] Although the federal government has paid the initial cleanup costs at many privately owned Superfund sites, the strict, joint, and several liability provisions of the act allow for recovery of expenditures when responsible parties can be identified.

The private waste management industry has seen a large degree of concentration since the early 1980s. It has not developed along the lines many anticipated in the early years of the program. At that time, most experts thought the future of waste management would be increased thermal treatment, such as incineration. Public opposition to thermal treatment and high prices resulted in much less capacity, and demand, than anticipated.

RCRA requires that all hazardous waste be treated, stored, and disposed of in permitted TSD facilities. In 1995, the latest date for which comprehensive data are available, 1,983 permitted TSD facilities handled hazardous waste. Of this total, 1,083 were storage facilities, leaving 900 treatment and disposal facilities. The largest fifty treatment and disposal facilities managed nearly 90% of the total volume of hazardous waste. In 1992, there were twenty-one major off-site commercial hazardous waste landfills and twenty major off-site commercial hazardous waste incinerators (one of which was in the same facility as one of the landfills).[16]

Most hazardous waste is treated and disposed of on site, with only 5% to 10% delivered off site for storage, treatment, and/or disposal. The on-site emphasis results from the fact that 95% of the total hazardous waste volume is wastewater, which is usually treated on site and then discharged to publicly owned treatment works. However, most generators dispose of at least some hazardous waste off site.

Table A-5 shows the HWM methods used at commercial and noncommercial permitted treatment and disposal (excluding storage) facilities in the United States. The "All Waste" column shows data for on-site and off-site treatment and disposal methods, while the "Waste Received from Off Site" column shows data for waste received from off site only. Of the 900 permitted treatment and disposal facilities, 732 receive waste from off site. Some facilities may use multiple methods and are, therefore, counted twice.

Most of the U.S. hazardous waste infrastructure was built before strict regulations on such facilities existed. Since that time, many of the poorest environmental performers have shut down due to more stringent requirements under RCRA and as a result of Superfund liability. Hazardous waste treatment and disposal facilities have had to come into compliance with new RCRA requirements over time. New facilities have been required to meet EPA's more stringent requirements.

Table A-5. Management Methods Used by Permitted Hazardous Waste Treatment and Disposal Facilities in the United States

Management Method	All Waste (Number of Facilities)	Waste Received from Off Site (Number of Facilities)
Aqueous treatment[a]	281	94
Land disposal[b]	123	51
Thermal treatment[c]	291	123
Recovery[d]	397	244
Other treatment and disposal[e]	467	197

Note: Facilities that are only storage units are excluded. Facilities with multiple handling methods may be counted more than once.

[a] Aqueous organic, inorganic, and combined (organic and inorganic) treatment units.

[b] Deep-well/underground injection, landfill, surface impoundment, and land treatment/application/ farming.

[c] Incineration and energy recovery (reuse as fuel).

[d] Fuels blending, metals recovery (or reuse), solvent recovery, and other recovery.

[e] Stabilization, sludge treatment, and other treatment and disposal.

Source: U.S. Environmental Protection Agency, National Biennial RCRA Hazardous Waste Report (Washington, D.C.: U.S. EPA, 1997), based on 1995 data.

As a result of the rise of public opposition to hazardous waste facility siting, there has been an almost universal halt in off-site facility construction since the mid-1980s. Most of the increase in treatment and disposal capacity, therefore, has been the result of increases in capacity at existing facilities. Major off-site treatment and disposal facilities are concentrated in a few geographical areas: Texas, Tennessee, Louisiana, Michigan, and California.

Canada

Much of Canada's environmental policymaking—including responsibility for hazardous waste management—is shared with its provincial governments. In general, provinces regulate hazardous waste activities that occur solely within their boundaries, while the federal government is responsible primarily for establishing national guidelines and regulating interprovincial and international transport of hazardous waste. Different provinces have taken quite different approaches to establishing hazardous waste infrastructure. Alberta and Manitoba, for example, have used public funds to construct HWM facilities, while Quebec, Ontario, and British Columbia have relied on the private sector to construct facilities. In this profile, we focus on one representative province for each approach: Alberta and Quebec.

Canada's federal definition of hazardous waste includes "those materials intended for disposal and which are potentially hazardous to human health and/or the environment due to their nature and quantity, and which require special handling techniques."[17] Lists, criteria, and definitions laid out in national legislation form the basis for a national definition of hazardous waste, but, historically, provinces had significant discretion in determining which wastes to regulate as hazardous. In the last three years, Canada has undertaken a major effort to harmonize hazardous waste lists, exemptions, and classification criteria across all provinces and territories.

Environment Canada, the federal environmental authority, estimates that the country produces around six million tons of hazardous waste annually.[18] Sixty percent is processed on site and 40% goes off site. Of that going off site, an annual total of approximately 1.8 million tons goes to treatment and disposal facilities. The country has five "principal" HWM facilities, defined as integrated treatment and disposal facilities processing over 50,000 tons of waste per year.[19] Ontario and Quebec produce approximately 75% of Canada's hazardous waste, with Saskatchewan, Alberta, and British Columbia producing approximately 7% each, and the remainder produced by Canada's other provinces and territories.

The following sections discuss hazardous waste regulatory development and HWM facilities in Canada in greater detail. Table A-6 presents a time line of important dates in the development of the country's HWM program.

Program Development

Federal and provincial legislation both influence hazardous waste management in Canada. The federal Transportation of Dangerous Goods Act (TDGA) was the first law to specifically regulate hazardous waste at the federal level.[20] It was passed in 1980 after public outcry about a train and tanker car accident involving liquid chlorine. TDGA's regulations, promulgated in 1985, established federal definitions for hazardous waste (leaving some discretion to provinces and territories) and established a system of manifests and notification for tracking interprovincial waste transfers.

Although TDGA primarily addressed the safe transport, rather than overall management, of hazardous waste, it provided an important incentive for provinces and territories to develop their own hazardous waste laws and to modernize disposal infrastructure. For example, Quebec promulgated its HWM regulations in 1985. According to personnel in Quebec's environment ministry, most of these regulations were implemented over the next three years.[21] Some analysts have charged, however, that regulation continued to be weak into the 1990s.[22]

Alberta also passed a comprehensive set of hazardous waste regulations in 1985, developed under the authority of a variety of laws, including the Hazardous Chemicals Act, the Clean Air Act, and the Clean Water Act.[23] By the end of the 1980s, Alberta's regulatory system was reportedly largely operational.[24]

The 1988 Canadian Environmental Protection Act (CEPA) led to new rules on the transboundary movement of hazardous waste. Under CEPA, Canada promulgated the Export and Import of Hazardous Waste Regulations (EIHWR) in 1992, which allowed the country to ratify the Basel Convention, an international treaty concerning the movement of hazardous waste between countries. EIHWR, along with the 1985 TDGA regulations and the various provincial waste management acts and regulations, currently forms the core legislation for controling hazardous waste in Canada.

In general, the 1990s have seen an increasing emphasis on recycling of hazardous waste rather than on disposal. The 1990s have also seen continued

Table A-6. Hazardous Waste Program Development in Canada

Year	Laws/Policy	Agency	Regulations	Facilities
1971		Department of Environment established (later became Environment Canada)		Mercier, Quebec, incinerator began operation
1980	Transportation of Dangerous Goods Act (TDGA)			
1982				Alberta Special Waste Management Corporation established
1983				Stablex TSD facility opened in Blainville, Quebec
1985			Alberta and Quebec established HWM regulations	
			TDGA regulations promulgated	
1987				Alberta Special Waste Treatment Center (ASWTC) built in Swan Hills, Alberta
1988	Canadian Environmental Protection Act			
1989			*Through 1992:* Federal HWM guidelines published by CCME	
1992	Alberta Environmental Protection and Enhancement Act		Export and Import of Hazardous Waste Regulations (amended in 1994)	
1996				ASWTC privatized
1997			Quebec established hazardous waste regulations (first major amendments to 1985 regulations)	

Notes: CCME, Canadian Council of Ministers of the Environment; HWM, hazardous waste management; TSD, treatment, storage, and disposal.

development of HWM policy in the provinces. Alberta passed the Environmental Protection and Enhancement Act in 1992, which consolidated HWM authority under one statute. In 1997, Quebec adopted the first major amendments to its 1985 hazardous waste regulations, aimed at more effectively targeting the largest firms responsible for generating, treating, and disposing of hazardous waste.

Overseeing federal environmental policy is Environment Canada, established as the Department of Environment in 1971. Each province also has its own environmental authority. In Alberta, for example, the main environmental regulatory body is Alberta Environmental Protection. In Quebec, it is the Ministère de l'Environnement et de la Faune. While the decentralization of authority to the provinces led to some differences in regulatory programs, current trends are toward increased harmonization. These efforts are led by the Canadian Council of Ministers of the Environment (CCME), which is composed of all provincial, territorial, and federal environmental ministers. CCME seeks to harmonize hazardous waste management and disposal policy throughout Canada by publishing consensus guidelines that outline minimum standards that should be met by provincial regulations. Many of these guidelines were developed between 1989 and 1992.

Hazardous Waste Management Facilities

In the late 1970s and early 1980s, the need for modern hazardous waste facilities began to be recognized throughout Canada. As already mentioned, provinces adopted two general models for financing the construction and operation of HWM facilities—one uses public financing and the other relies on the private sector. Environmental authorities attribute the different approaches to the different philosophy, geography, demographics, and industrial profile of each province.

The first approach, adopted by Alberta and Manitoba (and later British Columbia, although no facility was built), was to establish a government-owned crown corporation. Alberta's crown corporation, the Alberta Special Waste Management Corporation, was established in 1982 and helped construct the province's first (and only) large integrated treatment and disposal facility, located in Swan Hills, Alberta. The facility, known as the Alberta Special Waste Treatment Center (ASWTC), was built in 1987.

The crown corporation's private partner in constructing ASWTC was the firm Chem-Security. Chem-Security was responsible for day-to-day operations

and provided 60% of construction and operating costs. The remaining costs
were paid by the crown corporation, which also provided:

- a guaranteed return on investment to the private firm,
- partial responsibility for the collection and transfer of waste,
- full responsibility for postclosure activities,
- utilities and road access,
- research funding, and
- leasing of the publicly owned site to the private firm for a minimal fee.

In July 1996, Alberta privatized ASWTC, selling the facility to Bovar, Inc.
Although ASWTC played a large role in treatment and disposal in the province,
significant on-site treatment and disposal still exist, and a number of other private companies provide hazardous waste management and recycling services in
the province. Each year since the early 1990s, around three times as much hazardous waste (on average) has been recycled annually in Alberta as has been disposed of at the Swan Hills facility.[25]

The second model used in Canada for developing hazardous waste treatment and disposal capacity was to rely on private companies to build and operate facilities. The private-sector model is exemplified by Quebec, with its two
large HWM facilities. The first is a hazardous waste incinerator, which began
operation in 1971 in Mercier, Quebec. The second is the Stablex treatment, storage, and disposal facility in Blainville, which opened in early 1983.

Quebec's contribution to the Stablex facility was considerably less than
Alberta's contribution to Swan Hills. The province provided highway access,
secured the land for the facility from the federal government, and assisted with
the public consultation before approval of the project. Quebec's public funding
of hazardous waste management is currently limited to grants for research and
development, mainly in the area of recycling.

Malaysia

Over the last few decades Malaysia's economy has expanded rapidly and has
undergone a transition from agriculture to manufacturing. From the mid-1980s
to the mid-1990s, the growth rate of Malaysia's gross domestic product (GDP)
exceeded 8%. The size of the manufacturing sector has increased by 60% in ten
years. These trends have made hazardous waste management one of Malaysia's
most pressing environmental problems. The country began its process of developing a hazardous waste regulatory program in the early 1980s, coming out

with regulations in 1989. The government long sought to have the private sector finance and operate HWM facilities and only recently has the country's first modern TSD facility been built.

Hazardous waste is known in Malaysia as "scheduled waste." Regulations specify 107 categories of scheduled waste from nonspecific and specific sources. The Department of the Environment estimates that total scheduled waste production for 1996 was around 630,000 tons.[26] According to 1994 statistics, over half of the waste is produced by three sectors: metal finishing (28%), textiles (15%), and industrial gas production (14%).[27]

The following sections discuss hazardous waste regulatory development and HWM facilities in Malaysia in greater detail. Table A-7 presents a time line of important dates in the development of the country's HWM program.

Program Development

Malaysia's basic environmental legislation is the Environmental Quality Act (EQA) of 1974. It provides the legal basis for hazardous waste regulation in the country.

The first major investigation into hazardous waste problems in Malaysia was undertaken in 1981. The investigation's report recommended policies and guidelines for an HWM program, identified major sources of hazardous waste, named potential sites for a TSD facility, and established a list of hazardous wastes. The Task Force on Toxic and Hazardous Waste was created to implement the recommendations, leading to a draft of cradle-to-grave regulations for hazardous waste in 1984. In 1989, Malaysia promulgated its HWM regulations. Delays in developing regulations were attributed to a lack of trained personnel, a lack of facilities, and the difficulty in preparing schedules for toxic and hazardous wastes.[28] During the intervening period, a number of studies of hazardous waste generation and possible HWM facility locations were undertaken, and generators were encouraged to reduce waste production.

The 1989 regulations included the Environmental Quality (Scheduled Waste) Regulations and the Environmental Quality (Prescribed Premises) (Scheduled Wastes Treatment and Disposal Facilities) Regulations. They covered:

- definitions of toxic and hazardous waste;
- use, storage, handling, transport, labeling, and disposal requirements;
- monthly reporting of facility inventories;
- a notification and tracking system; and
- licensing of hazardous waste transporters and off-site TSD facilities.

Table A-7. Hazardous Waste Program Development in Malaysia

Year	Laws/Policy	Agency	Regulations	Facilities
1974	Environmental Quality Act (EQA)			
1975		Department of the Environment (DOE) established in Ministry of Science, Technology, and Environment		
1981	Report on first major investigation of hazardous waste	Pollution Control Division established in DOE		Study identified 12 potential TSD facility sites
		Regional DOE offices opened		
1984			Draft hazardous waste regulations developed	
1985	EQA Amendments			
1987				Feasibility study on TSD facility construction completed
1989			Hazardous waste regulations promulgated	15 proposals submitted for constructing TSD facilities
1992				Tentative agreement with government on Kualiti Alam facility
1996	EQA Amendments			
1998				Kualiti Alam facility opened

Note: TSD, treatment, storage, and disposal.

According to regulatory officials, most procedures are in place for implementing the 1989 hazardous waste regulations.[29] Amendments to EQA in 1996 and recent regulations have increased penalties for pollution violations, authorized a waste generation tax, and given the government authority to close down plants and require factory audits for firms that are violating the law. Hazardous waste regulations are currently being revised, and a new list of regulated wastes is being developed.

The main hazardous waste regulatory body in Malaysia is the Pollution Control Division of the Department of the Environment. The 1974 EQA authorized the creation of the Department of the Environment (DOE) within the Ministry of Science, Technology, and Environment, and the department was established in 1975. DOE is responsible for most of Malaysia's environmental regulatory implementation, development, and planning, as well as enforcement. The Pollution Control Division—which was specifically tasked with enforcing the 1974 EQA—was set up in 1981. In the same year, a number of regional DOE offices opened and were given regional enforcement and supervision responsibilities. DOE continues to decentralize its activities to the thirteen federal states. The department has a staff of roughly 600, one-third of which are inspectors.[30] In addition to DOE, the Environmental Quality Council, the Ministry of International Trade and Industry, and the Federation of Malaysian Manufacturers have significant influence and representation in environmental policymaking.

Enforcement officers visit registered facilities ("prescribed premises") at least once a year.[31] Malaysia has instituted penalties of fines and imprisonment for violators of hazardous waste laws. However, enforcement and compliance are regarded by some as slow, mainly due to a lack of staff and industry's "look-and-see" attitude.[32] Some enforcement does take place, however. For example, in 1997, authorities brought 275 cases against violators of various provisions of the 1974 EQA.[33] Newly authorized fines and the potential for violators' receiving prison terms are likely to get generators' attention.

Hazardous Waste Management Facilities

Although Malaysia's regulations governing the treatment and disposal of hazardous waste have been in place since 1989, the country has lacked (until recently) a large, modern treatment and disposal facility. Consistent with its privatization policies, Malaysia was committed to having the private sector construct, own, and operate a comprehensive waste management facility. In 1998, the firm Kualiti Alam opened Malaysia's first integrated and centralized hazardous waste treatment and disposal facility. Kualiti Alam is a private-sector joint venture between a Danish company, Danish Waste Treatment Services (25%), and two Malaysian firms, United Engineer Construction Sdn. Bhd. (50%) and Arab Malaysian Development Bhd. (25%). In addition to assuming the responsibility for full capital and operating costs, Kualiti Alam is responsible for collection and transportation of hazardous waste from generators to the facility.

The development of a modern TSD facility has been a long process in Malaysia. Starting with a feasibility study in 1987, Malaysia made a number of efforts through the late 1980s and early 1990s to attract companies to build a centralized TSD facility. In 1989, fifteen proposals were submitted to construct such a facility. In 1992, Kualiti Alam came to a tentative agreement with the government to build the facility, but it still took six years to site, construct, and open it.

The long time that elapsed between Malaysia's solicitation of bids to construct a facility in the late 1980s and the opening of the Kualiti Alam facility in 1998 can be blamed on a number of factors, including local opposition to facility siting. But, it was also a function of lengthy negotiations on the kinds of incentives the government would provide to private investors. One company, Waste Management International, withdrew from a project in the late 1980s, in part because the government would not grant revenue guarantees. After joining a private-sector joint venture in 1989, Chem-Security, Ltd., of Canada also pulled out of its project because of the Malaysian government's refusal to provide financial support.[34]

Kualiti Alam operates with no direct financial support or guarantees from the government. It has, however, been granted an exclusive license to build, operate, and maintain a centralized and integrated facility for fifteen years. The company has also received siting support and limited financial support through existing policies to promote high-technology industries.

The Kualiti Alam facility is located at Bukit Nanas near Port Dickson. The facility includes an incinerator with a capacity of 30,000 tons per year (TPY), a solidification facility (40,000 TPY), a physical/chemical treatment facility (5,000 TPY), and a landfill (30,000 TPY).[35]

Overall, it is probably too early to come to any conclusions about capacity and profitability at the Kualiti Alam facility because it opened so recently. Around the time of its opening, Kualiti Alam had reportedly signed contracts for 55,000 TPY of currently generated waste and 120,000 tons of stockpiled hazardous waste.[36] Three hundred generators have signed up to use the facility when it is fully operational, but only sixty are currently sending waste.[37]

The fact that regulations were promulgated in 1989 but a modern TSD facility was not available until 1998 has forced generators to seek other forms of waste management. To comply with the law, generators have had to store hazardous waste on site, minimize generation, or arrange for waste to be shipped abroad (under license from DOE). A few large facilities have incinerated waste on site.[38] Since the early 1990s, DOE has licensed a number of smaller "environmentally sound" off-site secure landfills and off-site pretreatment, recycling, and recovery facilities (159 were licensed in 1997 alone).[39] Illegal dumping, however, is thought to persist, particularly among small generators.[40]

Malaysia's private-sector approach to establishing HWM capacity is consistent with its overall privatization efforts—regarded as "the most ambitious and successful in the world."[41] Its current five-year plan (1996-2000) includes plans for significant privatization in the energy, transportation, and communications sectors. However, some financial support for hazardous waste infrastructure does exist. The Ministry of International Trade and Industry offers special tax incentives for factories to build on-site TSD facilities; few industries have taken advantage of the program. The Malaysian Industrial Development Authority (MIDA) provides tax holidays and allowances for industries that properly store, treat, and dispose of hazardous waste. MIDA also exempts the importation of pollution control equipment from import duties and sales taxes. Malaysia has instituted a number of market-based regulatory instruments, such as user fees and deposit-refund systems.

Hong Kong

Hong Kong is a special administrative region of the People's Republic of China. It reverted to Chinese sovereignty on July 1, 1997, after 156 years as a British colony. Many aspects of Hong Kong—such as its high per capita income and its sophisticated financial markets—make it similar to the developed countries discussed in this report. However, with regard to the timing and challenges of its HWM efforts (as well as its recent rapid growth), it is more similar to the developing countries we discuss and, therefore, has been grouped with them in our discussion.

Hong Kong has coordinated the implementation of hazardous waste regulations with the construction of a fully subsidized hazardous waste facility. It uses the term "chemical waste" for hazardous waste. Substances qualifying as chemical waste are specifically noted in lists (called "schedules"). Hong Kong's regulations define chemical waste as follows:

> Any substance or thing being scrap material, effluent, or an unwant-
> ed substance or by-product arising from the application of, or in the
> course of, any process or trade activity, and which contains any of
> the substances or chemicals specified in the schedule, would be
> regarded as chemical waste if such substance or chemical occurs in
> such form, quantity, or concentration so as to cause pollution or con-
> stitute a danger to health or risk of pollution to the environment.[42]

Hong Kong generates approximately 100,000 tons of hazardous waste per year.[43] Two aspects of waste generation are important in Hong Kong. First, most of the hazardous waste generators (over 90%) are small businesses. Second, Hong Kong has an intense population density: according to the 1996 census, 6.3 million people live in a small portion of an approximately 1,100-square-kilometer area. The result is that most of the small businesses (which are also small genera-tors) are located in multiuser high-rise buildings known as "flatted factories." This makes in-house waste treatment or even pretreatment facilities impracticable. The high percentage of small firms also means that many of Hong Kong's generators are unlikely to be able to pay full (or even partly subsidized) disposal prices.

The following sections discuss hazardous waste regulatory development and HWM facilities in Hong Kong in greater detail. Table A-8 presents a time line of important dates in the development of the country's HWM program.

Program Development

The 1980 Waste Disposal Ordinance (WDO) provides the foundation for man-aging all of Hong Kong's solid and hazardous waste. Although studies of toxic and hazardous wastes were undertaken as early as 1977, a concerted govern-ment effort to address hazardous waste was not outlined until the publication of the 1989 white paper *Pollution in Hong Kong: A Time to Act*, which catalyzed public attention regarding the problem of hazardous waste. Laws defining haz-ardous wastes and outlining their management were passed in 1991 as amend-ments to the WDO. Until that time, few efforts ensured that such wastes were not discharged directly to sewers and surface water.

The regulations implementing the 1991 WDO Amendments were promul-gated as the Waste Disposal (Chemical Waste) (General) Regulations in 1992

Table A-8. Hazardous Waste Program Development in Hong Kong

Year	Laws/Policy	Agency	Regulations	Facilities
1960s		EPCOM established		
1977	*Control of the Environment* report by EPCOM provided foundation for environmental policy	Environmental protection unit established in government Secretariat		
1980	Waste Disposal Ordinance (WDO) passed			
1981		Environmental Protection Agency established		
1982				Preliminary siting efforts began
1983			Development of "interim arrangements" for codisposal of hazardous waste with municipal waste	
1986		Environmental Protection Department established		
1987				Feasibility study for HWM facility on Tsing Yi Island began (CWTC)
1989	*Pollution in Hong Kong: A Time to Act* white paper spurred development of HWM program			
1991	Amendments to WDO outlined hazardous waste management			
1992			Waste Disposal (Chemical Waste) (General) Regulations promulgated	
1993				CWTC completed
1995				CWTC fees raised

Notes: CWTC, Chemical Waste Treatment Center; EPCOM, Environmental Pollution Advisory Committee; HWM, hazardous waste management.

and have been in force since 1993. The regulatory program includes provisions for:

•   registering hazardous waste generators;
•   notifying regulators about the disposal of certain chemical wastes (such as polychlorinated biphenyls);
•   outlining adequate packaging, labeling, and storage procedures for waste;
•   licensing waste collection and disposal facilities;
•   establishing a "trip ticket" system to track waste from production to disposal; and
•   making improper dumping subject to prosecution.

Procedures for registering generators; licensing transport, treatment, and disposal facilities; and conducting inspections appear to be well developed. To date, Hong Kong has registered 9,200 chemical waste producers.[44] Officials believe that this covers the vast majority of generators. (Because of the number of small generators, and their significant cumulative contribution to waste flows, there are no small-generator exemptions.) Additionally, officials have licensed seventy chemical waste collectors and thirty-four small chemical waste treatment and disposal facilities.[45]

To date, 277 fines (averaging HK$7,000, but up to HK$50,000) have been imposed for violations of chemical waste laws, 92 of them in 1997.[46] Most violations have concerned improper labeling and storage, failure to register as a waste producer, and failure to arrange for proper waste disposal. Overall, enforcement and publicity have raised awareness about proper waste disposal. Waste producers are managing wastes more safely, and some have introduced technologies for waste minimization.

The main regulatory body in Hong Kong is the Environmental Protection Department (EPD), an independent entity with all environmental protection and control responsibilities. The antecedents of EPD reach back to the 1960s with the establishment of the Environmental Pollution Advisory Committee (EPCOM). EPCOM directed the production of a 1977 report, *Control of the Environment,* that is credited as the foundation of Hong Kong's environmental administration. At the time the report was issued, a small environmental protection unit was set up in the government Secretariat to guide environmental policies and coordinate various agencies with environmental responsibilities.

In 1981, Hong Kong established the significantly larger Environmental Protection Agency. This agency was to draft regulations, design a system of "interim arrangements" for chemical waste management, collect information on waste flows, and recruit technical staff. In 1986, EPD came into its current form as a consolidation of the environmental staff and resources of six government departments, including its forerunner, the Environmental Protection Agency.

Hazardous Waste Management Facilities

When demand for proper waste disposal arose in the early 1980s, there were no facilities in Hong Kong to handle the waste. Preliminary siting efforts began around 1982. Recognizing the time it would likely take to develop a hazardous waste facility, Hong Kong developed what are referred to as "interim arrangements" in 1983; these arrangements principally involved codisposal of hazardous and municipal solid waste in landfills. The interim arrangements allowed regulatory staff to gain some experience in managing hazardous waste, served to identify waste producers, and generally raised awareness about the need for safe disposal. The quantity of hazardous waste processed through these interim arrangements was small, and most continued to be disposed of in an uncontrolled manner.

A feasibility study for establishing a facility on Hong Kong's Tsing Yi Island (the site of the current facility) began in 1987. The location was selected from five potential sites due to its proximity to waste-generating areas, transportation access, and distance from residential areas. Enviropace Limited won the contract to design, finance, construct, and operate a facility for fifteen years. Enviropace is a firm owned by a subsidiary of Waste Management International (Sun Hung Kai property group) and CITIC Pacific, an investment arm of the People's Republic of China. The facility, known as the Chemical Waste Treatment Center (CWTC), was completed in 1993.

The Hong Kong government paid the full capital and, at least initially, operating costs of the facility. The government repaid Enviropace for the facility's capital costs over a five-year period after the facility began operation. This financial arrangement put the initial financial risk of constructing the facility on Enviropace but transferred that responsibility to the government when the facility was successfully completed. (Construction costs were approximately US$155 million). As a result of this arrangement, the government now owns the facility and land, and operation of the facility will revert to the government after fifteen years or on default by the contractor.

In addition to reimbursing capital costs, the government guarantees a certain amount of operating revenue for Enviropace and pays a variable fee (depending on the type of waste) per ton for waste processed by the facility. The government then assumes responsibility for recovering costs from generators. Initially, disposal fees were completely subsidized by the government. In 1995, they were raised to 20% of operating costs. The government has also encouraged use of the facility by paying the full cost of containers and transportation and through its impressive efforts to register generators and enforce waste regulations.

The design capacity of the CWTC is 100,000 metric tons of waste annually. In the first years of operation, the CWTC received over 90,000 tons per year—a substantial percentage of total generation in Hong Kong. The facility includes high-temperature incineration, oil–water separation, physicochemical waste treatment, and residue solidification.

With the construction of the CWTC, and the phase-in of the hazardous waste regulations in 1993, most of Hong Kong's interim arrangements have been phased out. However, some wastes that cannot be treated at the central facility (such as asbestos), as well as residue from the CWTC, are still being codisposed of in municipal waste landfills. Currently, approximately 8,000 tons of waste per year are codisposed.[47]

A number of small private companies compete with the CWTC in the collection and disposal of hazardous waste. Existing legislation does not mandate the use of the CWTC, and these other firms are allowed to operate as long as they are licensed. There are currently seventy licensed chemical waste collection companies and thirty-four licensed waste treatment and disposal sites.[48] As of 1997, most of the licensed treatment and disposal facilities dealt with spent electroplating solutions and photofinishing waste.

The public financing of the hazardous waste facility in Hong Kong is typical of the government's more general approach of investing public resources in pollution control. Government investment in the CWTC is only one component of an expected investment in environmental infrastructure of US$2.5 billion between 1995 and 2000.[49] In fact, the precedent of free disposal had already been set: at least until 1995, Hong Kong provided all solid waste disposal at its landfills for free in order to control illegal dumping.

Thailand

Attention to hazardous waste management in Thailand has been spurred by rapid economic and industrial growth. Between 1990 and 1995, the country's gross domestic product exceeded 10% growth per year. Between 1969 and 1990, the number of registered facilities generating hazardous waste increased from around 600 to 50,000 (over 50% of them in Bangkok). More than 90% of the factories in Thailand are medium-sized and small.[50]

Hazardous waste infrastructure in Thailand has been developed with a mix of public and private investment. However, it is government policy to recover the full cost of constructing and operating facilities through disposal fees—regardless of how much public investment has been made. In this sense, Thailand's approach looks very much like the private-sector approaches of other countries. The charging of fees has not been matched by an effective regulatory system, however. Instead, the system lacks coherence and coordination among the various laws and agencies under which it operates.

Thailand has no consistent national definition of hazardous waste. According to the Hazardous Substances Act (1992), "hazardous waste" includes:

> explosive substances, flammable substances, oxidizing agents and peroxides, toxic substances, substances causing diseases, radioactive substances, mutant causing substances, corrosive substances, irritating substances and other substances, chemicals or otherwise which may cause injury to persons, animals, plants, properties, or environments.[51]

The Ministry of Industry, which has significant regulatory authority over industry, uses a different definition, with some substances overlapping the other definition and some not. In 1996, the overall quantity of hazardous waste generated in Thailand was estimated to be approximately 1.6 million tons.[52]

The following sections discuss hazardous waste regulatory development and HWM facilities in Thailand in greater detail. Table A-9 presents a time line of important dates in the development of the country's HWM program.

Program Development

Thailand has no integrated legal framework for hazardous waste management. Rather, authority is spread across at least three pieces of major legislation and their amendments:

- The 1967 Poisonous Substances Act gave the Ministries of Agriculture, Industry, and Public Health the authority to identify hazardous substances and to control their trade, storage, transport, usage, and disposal.
- The 1969 Factory Act gave the Ministry of Industry the authority to monitor and set requirements for the operation of factories, including their management of hazardous waste.

Table A-9. Hazardous Waste Program Development in Thailand

Year	Laws/Policy	Agency	Regulations	Facilities
1967	Poisonous Substances Act			
1969	Factory Act			
1975	National Environmental Quality Promotion and Conservation Act (NEQA)			
1978	NEQA Amendments	NEQA Amendments established National Environment Board (NEB)		
1983				Initiation of Samae Dam TSD facility
1988	*Through 1992*: National planning effort focused on need for comprehensive HWM system		*Through 1992*: Ministry of Industry promulgated variety of HWM announcements and regulations	Samae Dam TSD facility opened
1992	Enhancement and Conservation of National Environmental Quality Act Hazardous Substances Act Factory Act	NEB raised to subcabinet level Pollution Control Department created		
1994				General Environmental Conservation Company established
1997				Rayong TSD facility opened

Notes: HWM, hazardous waste management; TSD, treatment, storage, and disposal.

• The 1975 National Environmental Quality Promotion and Conservation Act was the first law dealing specifically with environmental quality generally. Its 1978 amendments created the National Environment Board to centralize the environmental authority that had been dispersed throughout the government.

In spite of this legislation, hazardous waste received little attention until various national planning efforts conducted between 1988 and 1992 identified the need for a coordinated system of hazardous waste management, treatment, and disposal. The Ministry of Industry promulgated a number of hazardous waste management announcements and regulations between 1988 and 1992.

Major legislative attention to hazardous waste came in 1992, with amendments to the three main environmental laws, which together significantly transformed the legal framework governing hazardous waste management and the consequences for noncompliance.

The main piece of 1992 legislation was the Enhancement and Conservation of National Environmental Quality Act, which replaced the National Environmental Quality Promotion and Conservation Act of 1975. The law's main contribution was to raise the National Environment Board (NEB) to a sub-cabinet level and to give it the authority to adopt and enforce pollution standards; this authority had previously been the purview of other government agencies. Of particular importance, the law gave NEB authority to enforce standards against industrial polluters, which had been the sole jurisdiction of the Ministry of Industry (and to a large extent remains so in practice). In the area of hazardous waste management, NEQA required the licensing of operators and inspectors of waste treatment and disposal facilities as well as record-keeping and reporting requirements. It also introduced provisions on liability for environmental damage. The law authorized provincial authorities to develop plans for procuring and constructing central waste disposal facilities or for promoting private-sector investment in waste management facilities.

The 1992 Hazardous Substances Act amended the 1967 Poisonous Substances Act. It is the primary law governing the manufacture, storage, transport, use, and disposal of hazardous substances. The law raised penalties for violating hazardous materials rules and established strict liability for accidents involving hazardous substances.

The 1992 Factory Act amended the 1969 Factory Act as the primary law governing operations within factories. The original law introduced requirements for constructing, operating, and expanding facilities. It created a registration system under which facilities must report, among other things, information about pollution control and treatment. It required licensing of new facilities by the Ministry of Industry and notification procedures for significant operating changes. The 1992 amendments extended the authority more explicitly to the management of hazardous wastes and pollution control. The amendments authorized the Ministry of Industry to limit pollution and waste discharges, to set standards for worker safety and accident prevention, and to reject licenses on environmental grounds. The law authorized fines up to US$80,000 and prison terms for plant managers operating without permits. It gave the Ministry of Industry the authority to shut down factories that did not comply with standards.

Thailand has not promulgated regulations implementing the 1992 laws. Although a ministerial announcement in 1988 adopted the U.S. definition of hazardous waste and another in 1990 adopted the Basel Convention definition, in practice what qualifies as hazardous waste remains ill defined. A national manifest system is currently being developed.

Just as the legal framework for hazardous waste management is spread among various laws, regulatory and enforcement responsibility is shared among a number of government bodies. This makes the institutional structure of hazardous waste regulation quite complex. Of particular relevance to toxic and hazardous waste are the entities with primarily environmental responsibilities—the National Environment Board and the Pollution Control Department within the Ministry of Science, Technology, and the Environment (MSTE)—and those with the power to regulate industry—the Ministry of Industry and the Industrial Estates Authority of Thailand.

The National Environment Board (NEB) has the primary responsibility for coordinating environmental policy and planning among government agencies. Before 1992, it was the primary office within MSTE dealing with hazardous waste. The board is chaired by Thailand's prime minister and is composed of the heads of relevant ministries, other government officials, and eight outside members, including at least four from private industry. NEB has the authority to set environmental quality standards, and it oversees the issuance of ministerial rules and regulations concerning the environment.

The Pollution Control Department was created within MSTE in 1992. It is tasked with developing the regulations to implement the 1992 Enhancement and Conservation of National Environmental Quality Act—including those relevant to the operation of waste facilities. MSTE is responsible for the inspection, control, and promotion of waste management facilities. It shares enforcement authority with the Ministry of Industry.

Under the authority of the 1969 Factory Act, the Ministry of Industry regulates the establishment and operation of factories, primarily through its Department of Industrial Works. The Department of Industrial Works registers all but the smallest (greater than seven workers) factories. The department has the authority to regulate how factories control, treat, and dispose of industrial wastes, which it does through its ability to issue (and revoke) three-year facility operating licenses. The Ministry of Industry has the ability to address what happens inside the plant, as opposed to the Pollution Control Department's jurisdiction over what comes out of it. The ministry has primary responsibility for developing regulations to implement the 1992 Factory Act, including those concerning the handling of waste. The Ministry of Industry can issue and enforce environmental quality norms, conduct inspections, and sanction violators; it also establishes and contracts out the operation of hazardous waste treatment and disposal facilities. The ministry has been hostile to command-and-control regulation, preferring to support programs for voluntary waste reduction.

The Industrial Estates Authority of Thailand (an independent agency under the Ministry of Industry) regulates industries within Thailand's twenty-three industrial estates. The estates are collections of industrial facilities, two-thirds of which are located in the Bangkok area. Each industry—whether under the control of the Ministry of Industry or the Industrial Estates Authority—is responsible for the treatment and disposal of its own waste. The relevant authority issues licenses, establishes standards, monitors pollution, enforces legislation, and provides treatment facilities for factories in its area.

Other ministries with some responsibility for hazardous waste management include the Ministry of Public Health and the Ministry of Agriculture and Cooperatives. Bangkok is its own administrative unit overseen by the Metropolitan Administration. It has its own hazardous waste regulations that apply only to Bangkok. The administration has its own staff, program, enforcement, and budget for environmental projects.

There is some question of whether the expertise exists within any of these agencies to implement a comprehensive hazardous waste management system. In a 1995 report on hazardous waste, the International Maritime Organization noted that a shortage of trained personnel, technology, and know-how was a significant constraint on proper waste management.[53] In 1996, the director of the

Environmental and Safety Control Division of the Industrial Estates Authority stated that there were not enough experienced personnel in Thailand to run hazardous waste treatment and disposal facilities should they be established in the four estates as planned at that time by the authority.[54] Thailand also lacks laboratory and analytical facilities and trained consultants.

Reportedly, many generators still do not comply with the waste management laws. In 1995, the International Maritime Organization attributed much of the blame to the regulatory system, stating that current legislation "lacks adequate provisions for ensuring compliance and implementation."[55] Problems included a lack of coordination among agencies, inadequate manpower, insufficient resources, and lack of enforcement. In 1995, the Thailand Development Research Institute (TDRI) summed up the problems as follows: "In the absence of strict enforcement of laws pertaining to hazardous waste disposal, voluntary compliance on the part of hazardous waste producers tends to be low as firms try to keep production costs to a minimum."[56] TDRI cited a "shortage of collective political will" as one of the primary reasons for inadequate waste management systems.

Although the main 1992 environmental laws included tough financial sanctions for polluters, there are significant barriers to enforcement. As of 1995, the ratio of regulatory staff to registered factories was one to one hundred, and the regulatory budget amounted to around US$76 per factory per year.[57] As of 1996, the Department of Industrial Works, under the authority of the 1992 Factory Act, had taken only thirty to forty enforcement actions against serious polluters.[58] The quality of information on waste generation is still poor, in large part because there is too little money and too few staff to collect it.[59]

Hazardous Waste Management Facilities

Government policy in Thailand is to award operation and management contracts to private-sector (or public–private) firms that are then responsible for recovering all costs (including capital costs) from generators for the collection, transportation, treatment, and disposal of hazardous waste.

The country's first hazardous waste treatment center was built at Samae Dam in the Bangkhuntien district, a western suburb of Bangkok. The project was initiated in 1983, and the facility became operational in 1988. The Thai government, through the Ministry of Industry, publicly financed facility construction. The government continues to own the facility but, consistent with its cost-recovery policies, grants five-year operating leases to private and public–private firms.

In addition to recovering variable costs through disposal fees, the operating firm pays a rental fee on the facility and a royalty fee based on the quantity of waste processed. Because the operating firm repays the capital cost of the facility, as well as pays the operating cost, it cannot subsidize disposal fees. As of mid-1996, General Environmental Conservation Company (GENCO) was operating the facility. GENCO was established in 1994 as a public–private joint venture between the Ministry of Industry (25%) and the private financial conglomerate GCN Holdings Company, Ltd. (75%).

The Samae Dam facility was developed to handle heavy metal-containing wastewater generated by about 200 small and medium-sized electroplating factories around Bangkok. Due to their size, these facilities had inadequate resources for on-site treatment, and their number made individual monitoring impracticable. These facilities include a chemical treatment plant, a chemical flocculation and sedimentation treatment plant, a chemical fixation process, and a facility for mixing hazardous sludge or solid waste into cement. The Samae Dam facility's total capacity is approximately 1,100 tons per day.[60] Initially, approximately 95% of waste delivered to the facility consisted of electroplating wastewater. An associated landfill, with 150,000 to 200,000 tons of capacity is located nearby (at Ratchaburi). Plans for the landfill were initiated in 1984, and the facility became operational in 1993.

In 1997, Thailand opened a second facility—a US$100 million integrated treatment and disposal facility in Rayong (Map Ta Put). GENCO financed the Rayong facility and contracted with Waste Management International for the project proposal, facility design, construction, and initial operation. GENCO has enjoyed some investment incentives from the government, including investment promotion policies from the Board of Investment (such as an exemption from 8% income tax and a tariff reduced to 10% from 20%–30%) and a guaranteed financial inflow and outflow, as well as protection under the Factory Act.[61]

The built capacity of the Rayong facility is about 550 tons per day.[62] Although this facility really supports only the industrial parks in its immediate area, all industrial estates are supposed to ship waste there. However, transportation costs are high, and distant industries are either storing waste or disposing of it on site.[63]

In spite of the existence of the Samae Dam facility and the new facility in Rayong, waste disposal problems in Thailand continue. As a result of the 1992 acts, generators without their own waste disposal facilities are required to transport waste to a central treatment plant and pay for treatment and disposal. However, in 1995, it was estimated that only 10%–20% of hazardous waste generated in Thailand was being treated at an approved facility.[64] The Samae Dam and Rayong facilities both run at only around 35% of capacity.[65] In the early 1990s, the Ministry of Industry encouraged all hazardous waste-generating factories to store waste on site. Many large producers still stockpile or treat wastes on site. Many small-scale waste producers have insufficient access to collection, treatment, and disposal facilities and either stockpile waste or dispose of it in waterways or municipal waste dumps.[66]

As of 1995, the Department of Industrial Works had plans to construct facilities at two additional sites around Bangkok: Chonburi and Saraburi. There have also been plans to construct hazardous waste disposal facilities at all of the country's industrial estates. However, those plans are on hold as a result of the current economic crisis.

Public opposition has been a major constraint on building additional facilities. For example, the construction of the hazardous waste treatment and disposal facility in Rayong was delayed in 1995 due to continued protests from the local population. In 1996, those protests led to the re-siting of the facility to a new location in an industrial complex in another part of Rayong.

The mostly private-sector approach to the facility in Rayong is more consistent with the Thai government's relatively hands-off approach to development than the public-sector approach at the Samae Dam facility. Most of Thailand's industrial development has been led by private, rather than public, investment. However, the government does fund or subsidize some public- and private-sector environmental projects. Within Thailand's Ministry of Finance is the Environmental Fund, which helps finance public and private pollution control. Overseen by the National Environment Board, the fund provides grants to government agencies and local administrations, primarily for municipal solid waste and wastewater treatment systems. The fund also provides low-interest loans to private firms for required disposal and treatment equipment and facilities. Before 1994, this fund had distributed US$200 million.

Indonesia

Indonesia is a collection of islands (approximately 13,700) that constitute most of the Malay Archipelago. Most of the country's industrial activity is concentrated on the island of Java, mainly around the capital city of Jakarta and other major cities—Surabaya, Bandung, and Semarang. In Indonesia, attention to hazardous waste management has been driven by rapid economic growth, averaging around 6% per year since the 1980s. At the same time, the manufacturing sector has expanded from around 13% of gross domestic product in the 1970s to around 33% in the 1990s. It is expected to rise to 45% in the next decade. These trends have increased attention to hazardous waste problems, particularly in Java, where the groundwater and most rivers are now seriously polluted with toxic and hazardous wastes.

Toxic and hazardous wastes are called "B3" wastes in Indonesia and are defined as wastes that "due to their characteristics and concentrations can indirectly or directly damage or pollute the environment and endanger human life."[67] In general, they are corrosive, flammable, reactive, explosive, or toxic. The definition includes infectious waste but excludes radioactive waste.

Estimates of hazardous waste generation in Indonesia vary widely. A country study by the United States–Asia Environmental Partnership reported that total hazardous waste generation in western Java and metropolitan Jakarta was approximately 2.2 million tons per year.[68] Waste Management International estimated a far smaller amount—approximately 200,000 to 250,000 tons per year.[69]

The following sections discuss hazardous waste regulatory development and HWM facilities in Indonesia in greater detail. Table A-10 presents a time line of important dates in the development of the country's HWM program.

Program Development

Indonesia's main environmental law is the 1982 Basic Provisions for the Management of the Living Environment (Law No. 4). It established the legal foundation for all environmental management, including hazardous waste management. Early regulations implemented by the Department of Health in 1983 and the Department of Industry in 1985 dealt with some aspects of toxic and hazardous waste, but they were not applied or enforced uniformly. The first draft of comprehensive toxic and hazardous waste regulations was prepared in 1987. A second draft was written in 1990 but was not passed.

The main government regulations concerning the management of hazardous waste were passed in 1994 (PP 19/1994) and 1995 (PP 12/1995). These came as part of a wave of environmental regulations and decrees between 1990 and 1994. The hazardous waste regulations and decrees establish a cradle-to-grave HWM system similar, at least in design, to that established in the United

Table A-10. Hazardous Waste Program Development in Indonesia

Year	Laws/Policy	Agency	Regulations	Facilities
1978		PPLH established		
1982	Basic Provisions for the Management of the Living Environment	PPLH became KLH		
1983				*Through 1985:* Various sites identified for HWM facilities
1986				*Through 1992:* Feasibility studies performed on various sites
1987			1st draft of hazardous waste regulations prepared	
1990			2nd draft of hazardous waste regulations prepared	
1991		BAPEDAL established		
1992				PPLI formed
1993		KLH became MLH		Memorandum of understanding between PPLI and government; facility construction began
1994			*And 1995:* Main hazardous waste regulations promulgated	PPLI facility opened
1997	New environmental management law gave BAPEDAL authority to administer penalities and use civilian inspectors			

Notes: BAPEDAL, Environmental Impact Management Agency; HWM, hazardous waste management; KLH, State Ministry for Population and Environment; MLH, State Ministry for the Environment; PPLH, Ministry of State Development Supervision and Environment; PPLI, PT Prasadah Pemunahan Limbah Industry.

States under the Resource Conservation and Recovery Act. They provide guidelines for:

* waste classification,
* requirements for TSD facilities,
* placarding and labeling requirements,
* facility permitting, and
* a manifest system.

Indonesia's main hazardous waste regulatory agency is the Environmental Impact Management Agency (BAPEDAL). BAPEDAL is responsible primarily for setting standards, implementing compliance programs, and enforcing regulations. It is the successor to a number of earlier government institutions with environmental responsibility. First was the Ministry of State Development Supervision and Environment (PPLH; formed in 1978), which had primary responsibility for environmental issues in Indonesia before the passage of the 1982 Basic Provisions law. After the 1982 law was passed, PPLH became the State Ministry for Population and Environment (KLH) and then, after a cabinet reorganization in 1993, the State Ministry for the Environment (MLH). Although MLH has primary responsibility for environmental policy formulation, it is not an implementing agency. The need for an agency to develop and enforce environmental standards led to the creation of BAPEDAL in 1991. An early challenge for BAPEDAL was sorting out its legal role and responsibilities relative to the roles of other national and provincial agencies.

BAPEDAL has reportedly had difficulty attracting well-trained and experienced staff. It has, however, grown from a staff of around thirty when it was created in 1991 to more than 300. Currently, BAPEDAL has fifteen to twenty people (including administrative support) responsible for the enforcement of hazardous waste regulations countrywide.[70] BAPEDAL performs some of its responsibilities through regional offices, but there is no formal delegation of authority to provincial governments for hazardous waste management.

The small size of BAPEDAL's hazardous waste staff and a political culture that favors voluntary compliance programs have limited the agency's ability to enforce HWM regulations among generators. Indonesia's weak civil and administrative legal systems further limit the ability to enforce regulations in a nonvoluntary manner. BAPEDAL does have the authority to assess fines and penalties against violators of HWM regulations. There have even been cases in which BAPEDAL has used its authority to issue stop-work orders for facilities that are out of compliance, although this level of confrontation is unusual.

Some procedures for implementing the hazardous waste regulations have been undertaken in a limited way. For example, the manifest system is functioning for the facilities that are already transporting waste to the country's single commercial TSD facility, PT Prasadah Pemunahan Limbah Industry (PPLI). However, BAPEDAL has focused mainly on various voluntary compliance programs—such as PROKASIH (surface water), Langit Biru ("Blue Skies" for air), Adipura ("Clean Cities" for solid waste), and PROPER (facility management)— that have been Indonesia's preferred approach to environmental compliance. One of these programs, PROPER, which publicly ranks the environmental performance of facilities, may be expanded to include HWM performance. In 1997, a new law superseding the 1982 Basic Provisions law was amended to give BAPEDAL police-type "investigator" authority, which may signal an increase in its power to regulate.

Hazardous Waste Management Facilities

Indonesia pursued a private-sector approach to developing HWM infrastructure. Consistent with this policy, construction of its first modern facility began in 1992 under the private joint venture company, PT Prasadah Pemunahan Limbah Industr (PPLI), near Jakarta.

The PPLI joint venture includes the Waste Management International subsidiary, Waste Management Indonesia, (70%); the Indonesian company, PT Bimantara Citra (25%); and BAPEDAL (5%). PPLI is responsible for all aspects of construction and operation of the facility. PPLI received only modest incentives from the Indonesian government to construct the facility. The incentives included the duty-free importation of materials and equipment, access to offshore financing at more attractive rates, and assurance that BAPEDAL would promulgate and enforce regulations. BAPEDAL's equity share in the joint venture was in exchange for land provided to the facility.

The PPLI facility opened in 1994. Before construction of the PPLI facility, Indonesia had identified a number of potential sites for hazardous waste facilities in the mid-1980s and undertaken a variety of feasibility studies for integrated hazardous waste treatment and disposal centers between 1986 and 1992.

The capacity of the PPLI facility is approximately 90,000 tons per year, but it has operated well under design capacity since it was built. In the first year of operation, the facility received only 10,000 tons of hazardous waste. It took two years to achieve its economic break-even point—a year later than expected. The facility has only recently reached a flow of 30,000 tons of hazardous waste per year, which is still a small fraction of the hazardous waste generated in the area. The financial crisis in Indonesia caused shipments to drop off in mid-1997.[71]

Feasibility studies have been conducted on additional facilities to serve other regions of Indonesia. The two highest-priority locations for future facilities are Surabaya (East Java) and East Kalimantan. An additional facility in the North Sumatra–Aceh region is also being considered.

Endnotes

1. Country data come from World Bank, *World Development Report, 1997* (Washington, D.C.: World Bank, 1997).

2. International Maritime Organization (IMO), "National Waste Management Profile for Germany," in *Global Waste Survey: Final Report* (London: IMO, 1995), p. 105.

3. Institute for Prospective Technological Studies (IPTS), *The Legal Definition of Waste and Its Impact on Waste Management in Europe* (Seville, Spain: European Commission Joint Research Center, IPTS, 1997), p. 21.

4. Umweltbundesamt, *Survey on Hazardous Waste Management in Germany* (Berlin: Umweltbundesamt, 1995), p. 20.

5. Joanne Linnerooth and Gary Davis, *Hazardous Waste Policy Management— Institutional Dimensions* (Laxenburg, Austria: International Institute for Applied Systems Analysis, 1984), p. 10.

6. F. Van Veen, "National Monitoring Systems for Hazardous Waste," in *Transfrontier Movements of Hazardous Waste* (Paris: Organisation for Economic Co-operation and Development, 1985), p. 85.

7. Organisation for Economic Co-operation and Development (OECD), *OECD Environmental Performance Reviews: Germany* (Paris: OECD, 1993), p. 57.

8. Linnerooth and Davis, *Hazardous Waste Policy Management*, p. 23.

9. Ibid., p. 19.

10. IPTS, *The Legal Definition of Waste*, p. 21.

11. "Net Profits up 43% at Kommunekemi in 1995," *Haznews* (July 1, 1996).

12. Mogens Moe, "Environmental Administration in Denmark," *Environment News*, No. 17 (1995), Sections 13.2.1 and 13.7.2 (available at http://www.mem.dk/mst/books/moe/).

13. Roger C. Dower, "Hazardous Waste," in *Public Policies for Environmental Protection*, edited by Paul R. Portney (Washington, D.C.: Resources for the Future, 1990), p. 165.

14. Organisation for Economic Co-operation and Development (OECD), *OECD Environmental Performance Reviews: United States* (Paris: OECD, 1996), p. 103.

15. Katherine N. Probst, Don Fullerton, Robert E. Litan, and Paul R. Portney, *Footing the Bill for Superfund Cleanups: Who Pays and How?* (Washington, D.C.: Brookings Institution and Resources for the Future, 1995).

16. William Gruber, "Hazardous Waste Landfills, 1993," *EI Digest* (April 1993); John Hanke, "Hazardous Waste Incineration, 1993," *EI Digest* (May 1993).

17. International Maritime Organization (IMO), "National Waste Management Profile for Canada," in *Global Waste Survey: Final Report* (London: IMO, 1995), p. 91.

18. Environment Canada, *Status Report on Hazardous Waste Management Facilities in Canada—1996* (Ottawa, Ontario: National Office of Pollution Prevention, February 1998), p. 8.

19. Ibid., p. 4.

20. Ibid., p. 15.

21. E-mail from Benoit Nadeau (Ministère de l'Environnement et de la Faune, Quebec) to T. Beierle, June 10, 1998.

22. Barry G. Rabe, *Beyond NIMBY: Hazardous Waste Siting in Canada and the United States* (Washington, D.C.: Brookings Institution, 1994), p. 125.

23. Environment Canada, *Status Report,* p. 21.

24. Communication between Antonio Fernandes (Alberta Environmental Protection) and T. Beierle, May 27, 1998.

25. Communication between Antonio Fernandes and T. Beierle, June 24, 1998.

26. E-mail from Lim Thian Leong (Center for Environmental Technologies, Malaysia) to T. Beierle, May 8, 1998.

27. "Hazardous Waste Management in Industrializing Countries," *Haznews,* No. 59 (February 1993).

28. Environmental Resources Management (ERM), *Public/Private Sector Cooperation in the Provision of Hazardous Waste Management Facilities* (London: ERM, 1994), p. B5.

29. E-mail from Ibrahim Shafii (Department of the Environment, Malaysia) to T. Beierle, May 14, 1998.

30. United States–Asia Environmental Partnership, "US-AEP Country Assessment: Malaysia" (available at http://www.usaep.org/country/malaysia.html).

31. E-mail from Lim Thian Leong to T. Beierle, May 8, 1998.

32. E-mail from Ibrahim Shafii to T. Beierle, June 16, 1998.

33. E-mail from Lim Thian Leong to T. Beierle, December 4, 1998.

34. ERM, *Public/Private Sector Cooperation,* pp.B7–B8.

35. E-mail from Arman Massoumi (Global Plasma Systems, Malaysia) to T. Beierle, May 22, 1998.

36. Ibid.

37. E-mail from Lim Thian Leong to T. Beierle, May 8, 1998.

38. Memorandum from Gordon R. Young (United States–Asia Environmental Partnership, Malaysia) to T. Beierle, May 27, 1998; ERM, *Public/Private Sector Cooperation*, p. B1.

39. E-mail from Lim Thian Leong to T. Beierle, December 9, 1998.

40. ERM, *Public/Private Sector Cooperation*, pp. B1 and B4.

41. United States–Asia Environmental Partnership, "US-AEP Country Assessment: Malaysia.

42. Environmental Protection Department, *A Guide to the Chemical Waste Control Scheme* (Hong Kong: Environmental Protection Department, 1992), pp. 5–6.

43. E-mail from R.C. Rootham (Local Control Office [Territory East], Environmental Protection Department) to K. Probst, December 5, 1998.

44. E-mail from K.K. Yung (Local Control Office [Territory East], Environmental Protection Department) to T. Beierle, April 25, 1998.

45. Ibid.

46. Ibid.

47. E-mail from Keith Gilges (United States–Asia Environmental Partnership, Hong Kong) to T. Beierle, June 9, 1998.

48. E-mail from K.K. Yung to T. Beierle, April 25, 1998.

49. Economist Intelligence Unit, "Hong Kong Industry: Environmental Report Card," *EIU ViewsWire* (September 4, 1997).

50. E-mail from Sombat Sae-Hae (Thailand Development Research Institute) to T. Beierle, May 27, 1998.

51. International Maritime Organization (IMO), "National Waste Management Profile for Thailand," in *Global Waste Survey: Final Report* (London: IMO, 1995), p. 183.

52. Ibid., reporting data from Division of Industrial Hazardous Waste Management.

53. IMO, "National Waste Management Profile for Thailand," p. 183.

54. "Chemcontrol Symposium II: Developing Markets," *Haznews,* No. 104 (November 1996).

55. IMO, "National Waste Management Profile for Thailand," p. 183.

56. Thailand Development Research Institute (TDRI), *The Monitoring and Control of Industrial Hazardous Waste: Hazardous Waste Management in Thailand* (Bangkok: TDRI, 1995), p. 1.

57. IMO, "National Waste Management Profile for Thailand," p. 188.

58. United States–Asia Environmental Partnership, "US-AEP Country Assessment: Thailand" (available at http://www.usaep.org/country/thailand.html).

59. Communication between Satit Sanongphan (United States–Asia Environmental Partnership, Thailand) and T. Beierle, June 2, 1998.

60. Letter from Suriya Supatanasinkasem (General Environmental Conservation Company) to T. Beierle, January 14, 1999.

61. E-mail from Sombat Sae-Hae to T. Beierle, June 12, 1998.

62. Letter from Suriya Supatanasinkasem to T. Beierle, January 14, 1999.

63. Communication between Satit Sanongphan and T. Beierle, June 2, 1998.

64. TDRI, *The Monitoring and Control of Industrial Hazardous Waste,* p. 18.

65. Letter from Suriya Supatanasinkasem to T. Beierle, January 14, 1999.

66. IMO, "National Waste Management Profile for Thailand," p. 183.

67. ERM, *Public/Private Sector Cooperation,* p. A5.

68. United States–Asia Environmental Partnership, "US-AEP Country Assessment: Indonesia" (available at http://www.usaep.org/country/indonesia.html).

69. Patrick Heininger, "Solving the Hazardous Waste Problem in Developing Countries," unpublished manuscript (Singapore: Waste Management International, January 1998).

70. Fax from Dan Millison (Ecology and Environment, Inc.) to T. Beierle, April 9, 1998.

71. Heininger, "Solving the Hazardous Waste Problem in Developing Countries."

Appendix B: Project Contacts and Reviewers

In addition to reviewing secondary sources to develop the summaries of each country's regulatory programs and hazardous waste management facilities, we consulted individuals about each of the eight countries studied. People who provided us with information and sources for the country profiles have an asterisk (*) by their names. Once we developed draft country profiles, we sent them out for review. Individuals who provided comments on these written profiles have a dagger (†) by their names.

We also distributed a preliminary draft of the complete report to a number of people for their review and comment.

Country Profiles

Canada

Patricia Armitage,* Industrial Waste and Wastewater Branch, Alberta Environmental Protection

Antonio Fernandes,*† Air and Water Approvals Division, Alberta Environmental Protection

Alan Jenkins,* Ontario Department of the Environment

John Myslicki,† Transboundary Movement Division, Environmental Protection Service, Environment Canada

Benoit Nadeau,*† Service des Materiès Dangereuse, Ministère de l'Environnement et de le Faune, Quebec
Alex Salewski,* Waste Reduction Branch, Ministry of the Environment and Energy, Ontario
Harry Vogt,* Technical Services and Special Waste, Ministry of Environment, Lands, and Parks, British Columbia

Denmark

Marianne Christensen,* Miljostyrelsen (Danish Environmental Agency)
Sibylle Grohs,* DGXI, European Commission
Helle Peterson,* DGXI, European Commission
Suzanne Arup Veltze,* DAKOFA (Danish Waste Management Association)

Germany

Frank Reiner Billigman,*† Bundesverband der Deutschen Entsorgungs Wirtschaft (BDE)
Hans-Joachim Schulz-Ellermann,*† Bundesverband der Deutschen Entsorgungs Wirtschaft (BDE)
Gerhard Smetana,† Umweltbundesamt (Federal Environmental Agency)
Ella Stengler,*† Arbeitgemeinschaft der Sonderabfall-entsorgungs-gesellschaften der Lander—AGS (Association of State-Owned Companies for the Disposal of Hazardous Waste)
Joachim Wuttke,† Umweltbundesamt (Federal Environmental Agency)

Hong Kong

K.K. Yung,* Environmental Protection Department
Keith Gilges,† United States–Asia Environmental Partnership, Hong Kong
R.C. Rootham,* Local Control Office (Territory East), Environmental Protection Department
Albert Leung,* United States–Asia Environmental Partnership, Hong Kong

Indonesia

Patrick Heininger,* Waste Management International (also provided information on Malaysia, Thailand, and Hong Kong)
Dan Millison,*† Ecology and Environment (also reviewed Malaysia, Thailand, and Hong Kong)

Malaysia

Michael Hansen,* Kualiti Alam, Inc.
Lim Thian Leong,*† Center for Environmental Technologies (CETEC)
Arman Massoumi,* Global Plasma Systems, Corp.
Ibrahim Shafii,*† Department of the Environment
Gordon Young,*† United States–Asia Environmental Partnership, Malaysia

Thailand

Nisakorn Kositratna,* Division of Hazardous Substance and Waste Management, Pollution Control Department
Sombat Sae-Hae,*† Thailand Development Research Institute
Satit Sanongphan,*† United States–Asia Environmental Partnership, Thailand
Suriya Supatanasinkasem,* General Environmental Conservation Company

United States

Julie Gourley,† International and Special Projects Branch, Office of Solid Waste, U.S. Environmental Protection Agency (also reviewed Canada and Hong Kong)
Matt Hale,† Office of Solid Waste, U.S. Environmental Protection Agency
Joel S. Hirschhorn,† Hirschhorn and Associates

All Europe

Chris Clarke,*† Independent Consultant, former editor of *Financial Times' Environmental Liability Report*

Report Review

A number of people reviewed a copy of the preliminary draft of this report and gave us very helpful comments.

Ruth Greenspan Bell, Resources for the Future
Dave Campbell, National Office of Pollution Prevention, Environment Canada
Chris Clarke, Independent Consultant, former editor of *Financial Times' Environmental Liability Report*
Terry Davies, Resources for the Future
Keith Gilges, United States–Asia Environmental Partnership, Hong Kong

Matt Hale, Office of Solid Waste, U.S. Environmental Protection Agency
Alan Krupnick, Resources for the Future
Allen Kneese, Resources for the Future
Lim Thian Leong, Center for Environmental Technologies (CETEC), Malaysia.
    Comments also included input from Tan Meng Leng, former director gen-
    eral of the Department of the Environment, Malaysia; Damian Lim, free-
    lance environmental consultant, Malaysia; and Goh Kiam Seng, CETEC.
Dan Millison, Ecology and Environment, United States
Bekir Onursal, World Bank
R.C. Rootham, Local Control Office (Territory East), Environmental Protection
    Department, Hong Kong
Sombat Sae-Hae, Thailand Development Research Institute
Hans-Joachim Schulz-Ellermann, Bundesverband der Deutschen Entsorgungs
    Wirtschaft, Germany
Ellen Spitalnik, U.S. Environmental Protection Agency
Tom Walton, World Bank
Fumikazu Yoshida, Faculty of Economics, Hokkaido University, Japan

Resources for the Future is an independent, nonprofit organization engaged in research and public education with issues concerning natural resources and the environment. Established in 1952, RFF provides knowledge that will help people to make better decisions about the conservation and use of such resources and the preservation of environmental quality.

RFF has pioneered the extension and sharpening of methods of economic analysis to meet the special needs of the fields of natural resources and the environment. Its scholars analyze issues involving forests, water, energy, minerals, transportation, sustainable development, and air pollution. They also examine, from the perspectives of economics and other disciplines, such topics as government regulation, risk, ecosystems and biodiversity, climate, Superfund, technology, and outer space.

Through the work of its scholars, RFF provides independent analysis to decisionmakers and the public. It publishes the findings of their research as books and in other formats, and communicates their work through conferences, seminars, workshops, and briefings. In serving as a source of new ideas and as an honest broker on matters of policy and governance, RFF is committed to elevating the public debate about natural resources and the environment.